UNIVERSITY OF STRATHCL'

30125 00814785

D1761285

It should not b............... each generation to rediscover principles of process safety which the generation before discovered. We must learn from the experience of others rather than learn the hard way. We must pass on to the next generation a record of what we have learned.

Books are to be returned on or before the last date below.

mmun

a
in 1961
erican

eived,
le series
le 1960s
mented
Safety

LIBREX —

UNIVERSITY OF
STRATHCLYDE LIBRARIES

safety

sharingtheexperience

improving the way lessons are learned
through people, process and technology

ANDERSONIAN LIBRARY

WITHDRAWN
FROM
LIBRARY STOCK

UNIVERSITY OF STRATHCLYDE

BP Process Safety Series

Hazards of Oil Refining Distillation Units

A collection of booklets describing hazards and how to manage them

bp

IChemE

Institution of Chemical Engineers

This booklet is intended as a safety supplement to operator training courses, operating manuals, and operating procedures. It is provided to help the reader better understand the 'why' of safe operating practices and procedures in our plants. Important engineering design features are included. However, technical advances and other changes made after its publication, while generally not affecting principles, could affect some suggestions made herein. The reader is encouraged to examine such advances and changes when selecting and implementing practices and procedures at his/her facility.

While the information in this booklet is intended to increase the store-house of knowledge in safe operations, it is important for the reader to recognize that this material is generic in nature, that it is not unit specific, and, accordingly, that its contents may not be subject to literal application. Instead, as noted above, it is supplemental information for use in already established training programmes; and it should not be treated as a substitute for otherwise applicable operator training courses, operating manuals or operating procedures. The advice in this booklet is a matter of opinion only and should not be construed as a representation or statement of any kind as to the effect of following such advice and no responsibility for the use of it can be assumed by BP.

This disclaimer shall have effect only to the extent permitted by any applicable law.

Queries and suggestions regarding the technical content of this booklet should be addressed to Frédéric Gil, BP, Chertsey Road, Sunbury on Thames, TW16 7LN, UK. E-mail: gilf@bp.com

All rights reserved. No part of this publication may be reproduced, stored in a retrieval system, or transmitted, in any form or by any means, electronic, mechanical, photocopying, recording or otherwise, without the prior permission of the publisher.

UNIVERSITY OF STRATHCLYDE
19 MAY 2008
UNIVERSITY LIBRARY

Published by
Institution of Chemical Engineers (IChemE)
Davis Building
165–189 Railway Terrace
Rugby, CV21 3HQ, UK

IChemE is a Registered Charity in England and Wales
Offices in Rugby (UK), London (UK), Melbourne (Australia) and Kuala Lumpur (Malaysia)

© 2008 BP International Limited

ISBN-13: 978 0 85295 522 2

First edition 2008

Typeset by Techset Composition Limited, Salisbury, UK
Printed by Henry Ling, Dorchester, UK

665·532
HAZ

Foreword

Crude and Vacuum distillations are the first processing step of oil refining. These units are normally running with fairly large hydrocarbon inventories, at high throughputs and high temperatures up to 360–434°C (680–813°F), well above the auto-ignition temperatures of hydrocarbons. Hydrogen Sulphide and LPG are also present. While distillation is a well known and understood process, numerous incidents have occurred in the past and still happen regularly in the industry.

This booklet is intended for those operators, engineers and technicians working on Crude and Vacuum Distillation Units in order to raise awareness around this activity, and to promote the adoption of safe designs and practices to avoid the occurrence of such incidents.

I strongly recommend you take the time to read this book carefully. The usefulness of this booklet is not limited to operating people; there are many useful applications for the maintenance, design and construction of facilities.

Please feel free to share your experience with others since this is one of the most effective means of communicating lessons learned and avoiding safety incidents in the future.

JJ Gomez, Vice President Refining Safety & Operational Excellence

Acknowledgements

The co-operation of the following in providing data and illustrations for this edition is gratefully acknowledged:

- BP Refining Process Safety Community
- BP Separations Technical Community
- John Atherton
- John Bond, IChemE Loss Prevention Bulletin editorial panel member.

How to use this booklet

This booklet is intended as a safety supplement to operator training courses, operating manuals, and operating procedures. It is provided to help the reader better understand the 'why' of safe operating practices and procedures in Crude and Vacuum Distillation Units (CDUs and VDUs). This booklet should be read in conjunction with the BP Process Safety Series Booklets, in particular:

- Hazards of Water
- Hazards of Air and Oxygen
- Hazards of Steam
- Safe Furnace and Boiler Firing
- Hazards of Trapped Pressure and Vacuum
- Safe Handling of Light Ends
- Confined Space Entry
- Hazardous Substances in Refineries
- Safe Ups and Downs for Process Units
- Hazards of Electricity and Static Electricity
- Hazards of Nitrogen and Catalyst Handling

See also Incidents that define Process Safety: learning from past major industrial accidents (ISBN 978 0 47012 204 4)

> Text contained in YELLOW boxes summarizes incidents that are relevant to the adjoining script (numbers in parenthesis relate to extracts from various incident reports listed in Chapter 9).

> Text contained in BLUE boxes gives advice on where further guidance can be obtained.

> Text contained in ROSE boxes gives guidance to operators on safety and operational issues and the appropriate courses of action they can follow.

In addition to the 100 or so brief incidents described, Chapter 6 gives detailed descriptions of six serious incidents that occurred on Crude and Vacuum Distillation Units (CDU/VDUs).

A self test questionnaire is included in Chapter 7, with answers at the end.

Contents

1

Introduction

1.1　Process description

Crude oil enters the refinery by pipeline or tanker and is initially processed in the crude atmospheric and vacuum distillation units (referred to as CDU and VDU) to separate it into a number of wide-boiling fractions.

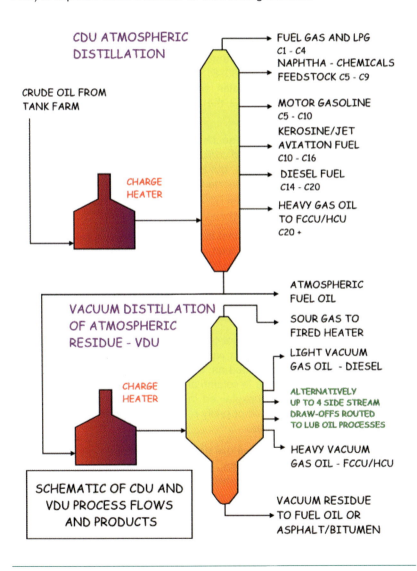

CDU ATMOSPHERIC DISTILLATION

CRUDE OIL FROM TANK FARM

CHARGE HEATER

FUEL GAS AND LPG
C1 - C4
NAPHTHA - CHEMICALS FEEDSTOCK C5 - C9

MOTOR GASOLINE
C5 - C10

KEROSINE/JET AVIATION FUEL
C10 - C16

DIESEL FUEL
C14 - C20

HEAVY GAS OIL TO FCCU/HCU
C20 +

ATMOSPHERIC FUEL OIL

VACUUM DISTILLATION OF ATMOSPHERIC RESIDUE - VDU

CHARGE HEATER

SOUR GAS TO FIRED HEATER

LIGHT VACUUM GAS OIL - DIESEL

ALTERNATIVELY UP TO 4 SIDE STREAM DRAW-OFFS ROUTED TO LUB OIL PROCESSES

HEAVY VACUUM GAS OIL - FCCU/HCU

VACUUM RESIDUE TO FUEL OIL OR ASPHALT/BITUMEN

SCHEMATIC OF CDU AND VDU PROCESS FLOWS AND PRODUCTS

Many different types of crude are processed making products including gas, gasoline, heating and industrial fuel oils, aviation and diesel fuels, lube oils, asphalt and coke. The crude oil is heated to between 343 to 382°C (650 to 720°F) before being charged to an atmospheric tower where the initial separation into different streams by boiling range is made. The residue from this tower is usually further heated to between 400–434°C (752–813°F), depending upon the type of operation required, before entering a vacuum tower where further heavier products are recovered. Most of the streams produced in the crude unit are further processed as feed for other refinery units where further separation and conversions are made.

1.2 Summary of main hazards

The CDU/VDU is the primary unit on the refinery, and is likely to be the oldest and most debottlenecked unit. These units are normally running with fairly large hydrocarbon inventories, at high throughputs and high temperatures up to 360–434°C (680–813°F). For the heavier product fractions (kerosene, gas oils, vacuum distillates, and all residues), the column operating temperatures are generally above the auto-ignition temperatures of these hydrocarbons.

The feed qualities vary widely with sour/sweet crudes, and crude oils which are acidic and can result in accelerated corrosion rates. It is likely that the unit was designed for a specific crude mix, but is most probably operated on a wide range of new crudes. It is very important that the unit design has the correct metallurgy for the range of crudes typically processed. Hydrogen Sulphide (H_2S) is also a permanent hazard in these units.

Good desalter operation is essential in order to minimize corrosion in CDU overheads due to chlorides in the feed, and allow acceptable brine quality for waste water treatment.

During normal operation, the main safety concerns for crude and vacuum units relate to possible loss of hydrocarbon containment due to materials corrosion, seal and flange leaks, and maintenance operations.

The reverse of loss of containment to the atmosphere is air in-leakage on VDUs. Small air in-leakages can result in combustion within a fuel-rich environment, sometimes described as 'cool flames', which can result in hot spots and mechanical damage to column and system internals. Air in-leakage in VDUs may result in an unexplained increase in sour gas flows. Large amounts of air in-leakage could possibly result in a flammable mixture and a possible ignition of the sour gas.

There are large fired heaters on these units, often dual fired with oil and gas, which have their own inherent hazards, and provide a ready source of ignition in the event of a major loss of containment of hydrocarbon.

Crude oil delivered to the refinery contains some water, which can enter the unit as slugs of water if tank farm operations are not properly controlled, or as entrained water which the desalter is designed to remove. Slugs of water can cause a major plant upset resulting in damage to column internals and possible overpressure of process equipment.

Crude oil also contains a wide range of organic and inorganic materials that can, either by themselves or through reaction with other process materials or the materials of construction of the plant, create aggressive corrosion sites.

Within the oil refining industry there have been many incidents of pyrophoric scale fires inside distillation columns during shutdown. This is a very significant safety hazard—particularly inside columns containing packed internals.

During start-up, operation of the unit under abnormal process conditions is a concern, particularly for dry operation of pumps and operation of the unit with liquids in normally vapour-only lines. Slug flow in overhead vapour lines is a serious hazard that can result in damage to overhead lines and pipe supports. Where liquid slugs can accumulate above pressure relief valves sized for vapour flow (as happened at Texas City isomerization unit incident in 2005) the release of liquid into large bore piping systems designed to handle vapour flows creates hazards of mechanical overloading of piping and supports, overfilling of flare knockout drums.

1.3 When incidents occur

Accidents occur throughout the life of a process plant. The most vulnerable times are during start-up and shutdown. An analysis of the numbers of incidents on CDUs and VDUs reported through an oil company safety bulletin since 1971 shows the following breakdown:

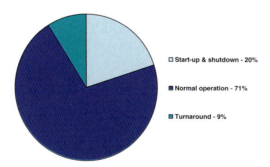

☐ Start-up & shutdown - 20%

■ Normal operation - 71%

■ Turnaround - 9%

Although the largest *number* of accidents are reported when the unit is in normal operation, including maintenance carried out during that period, this accounts for around 95% of the time. The time when the plant is being started-up and shutdown is much less, typically about one month every four years. Turnarounds take up the balance, around 3% of the time, say six weeks every four years. If the above were *normalized with respect to time*, the result would be:

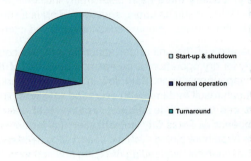

□ Start-up & shutdown

■ Normal operation

■ Turnaround

This clearly shows that start-up and shutdown are the most vulnerable times, turnarounds come a clear second, with normal operation in third place.

2

Chemical hazards

BP Process Safety Booklets cover the Hazards of Water, the Hazards of Air and Oxygen, the Hazards of Steam, and the handling of Hazardous Substances in Refineries.

2.1 Hydrocarbons

Crude oil is made up of a multitude of hydrocarbons ranging from methane to asphalt (bitumen (asphalt)). These are either flammable—they exist above their flash points at normal ambient temperatures and atmospheric pressure, or combustible—they need to be heated up to their flash point in order to be ignited by a sufficiently energized ignition source. In terms of flammability there are two major hazards:

Light hydrocarbons

For light hydrocarbons, typically C1 to C6, released from containment at normal ambient temperatures and heavier hydrocarbons heated above their flash points or released as a jet or mist. Hydrocarbons released from containment will readily form a cloud of flammable material that can travel long distances before reaching a source of ignition, such as fired heaters, hot work, vehicles, or unprotected electrical equipment. When this happens, the cloud burns back to the source of the release as a flash fire, during which time it can generate damaging overpressures that increase with the amount of plant and equipment congestion that the flame front has to travel through. The release then burns as a jet fire at the source of release, causing rapid failure of piping, process equipment or structural steel members that the flame impinges on. This is the mechanism that occurred at Flixborough in 1974 and Buncefield in 2005, for example. Critical properties of flammable hydrocarbons are flammable range in air and flash point as given below:

Material	Flammable range % volume in air	
	LEL	UFL
Methane	5.0	15.0
Propane	2.1	9.5
Butane	1.6	8.5
Gasoline	1.4	7.6
Naphtha	1.1	5.9
Kerosene	0.7	5.0
Hydrogen Sulphide	4.3	46.0

Note: The heavier the hydrocarbon, the smaller the LEL (lower flammable limit) and flammable range = [UFL (upper flammable limit) – LEL]

Typical boiling ranges and flash points for hydrocarbons that are normally liquid at ambient temperatures are given in the table below.

Product	Boiling range		Flash point	
	Deg C	Deg F	Deg C	Deg F
Gasoline	0–200	32–392	−43	−45
Naphtha	100–177	212–350	−2 to +10	28 to 50
Kerosene	151–301	304–574	43 to 72	109 to 162
Gas oil	260–371	500–700	>66	>151
Atmospheric residue	>350	>662	>100	>212

Notes:

- *The lowest temperature at which enough vapours are given off to form a flammable mixture of vapour and air immediately above the liquid surface is the **flash point** of the liquid fuel.*

- *These are typical values showing the boiling range from the IBP (initial boiling point) through to the FBP (final boiling point).*

- *Source (both tables): Institute of Petroleum (UK) Refining Safety Code 1983*

> **WARNING: Some heavy oils, particularly residues that have been processed at temperatures high enough for thermal cracking to occur, can contain small amounts of light hydrocarbons (such as methane, H₂S)** that may come out of solution in storage tanks giving rise to a fire and explosion hazard at far lower temperatures than the flash point. *Consult appropriate Material Safety Data Sheets (MSDSs).*

Heavier hydrocarbons

For heavier hydrocarbons that do not typically form a flammable vapour cloud on release from containment the major hazards are Auto-Ignition Temperature (AIT) and pool fire. Heavier hydrocarbons typically have low AITs, i.e. they can ignite purely through their internal heat at relatively low temperatures without having to find an external source of ignition. The chart below shows typical values of AITs. However, these should be treated with caution as the method of determination may not always replicate the conditions found in the field. Experience shows that hydrocarbon products in the range kerosene to residues almost always auto-ignite when released from containment at the temperatures they are produced at in CDU/VDU main columns. All hydrocarbon materials released at temperatures below their AIT can be ignited if they contact hot equipment and piping, or are released as mists or aerosols which can be ignited at far lower temperatures.

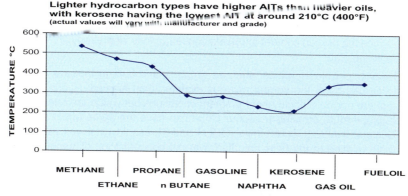

Lighter hydrocarbon types have higher AITs than heavier oils, with kerosene having the lowest AIT at around 210°C (400°F) (actual values will vary with manufacturer and grade)

TEMPERATURE °C

METHANE   PROPANE   GASOLINE   KEROSENE   FUELOIL

ETHANE   n BUTANE   NAPHTHA   GAS OIL

Source: IP Model Code of Safe Practice Part 3 – Refining Safety Code, 1981

Pool fires exhibit lower thermal energy than jet fires, but over a much larger area, causing failure of piping, process equipment and structural steel that lies within the thermal plume.

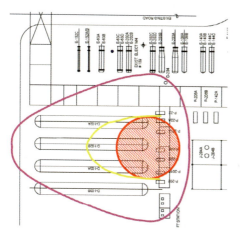

Example of a pool fire modelization for a fire hazard analysis of a CDU. Here a 100 mm (4 inches) hole in the line to a CDU desalter leads to a release generating a pool fire (red circle) with flame drag (yellow) and 37 kW/m² heat radiation contour (purple).

These calculations can be used to determine plant layout, fixed and mobile emergency resources and equipment, passive fire protection requirements and to check adequacy of emergency isolation valves.

A major pool fire resulted in the collapse of a Vacuum Distillation Unit main column across the pipetrack, probably because of lack of passive fire protection of the column skirt.

All hydrocarbons create occupational health and hygiene hazards

- Light gases, such as methane and ethane are asphyxiates, i.e. they can reduce the oxygen content to unsafe concentrations where they are allowed to accumulate.

- Heavier petroleum gases, such as propane and butane that are sold as LPG, can harm the central nervous system when inhaled. Hydrocarbons that are normally liquid under ambient conditions but which are released at a temperature creating a vapour cloud can have the same effect. Typical Occupational Exposure Limits (OELs) and Threshold Limit Values (TLVs) (both relating to the time weighted average concentration of vapour in air that workers can be continuously exposed to without harm throughout an 8-hour working day) are: propane and butane—600 ppm; pentane and n-heptane—500 ppm; naphtha—100 ppm.

- All liquid hydrocarbons have a solvent effect on skin, which can result in dermatitis if exposures are prolonged.

- Heavier hydrocarbons can be carcinogenic to skin if there is prolonged exposure, especially those that have been thermally cracked by high processing temperatures, such as the atmospheric and vacuum gas oils and residues.

- Aromatics contained within crude oil and refined products are highly likely to be carcinogenic.

- Those hydrocarbons that are liquefied by pressure alone, such as propane and butane, can cause freezing burns when exposed to bare skin.

- Thermal burns can result from contact with almost any process stream on a CDU/VDU except for those that have been cooled to ambient temperatures.

> **WARNING: When handling any hydrocarbon or chemical substance/ product make sure you comply fully with the Safety Warnings and Signs which define the hazards to you and your fellow workers.**

> **WARNING: If in doubt consult the material MSDS (Material Safety Data Sheet) or your local Occupational Hygiene (in the USA Industrial Hygiene) Advisor. *Do not proceed without being certain you are fully protected from the material hazard.***

2.2 Inorganic and other materials imported with crude oil

Sulphur

By far the most common inorganic material normally found in crude oil is sulphur, which comes in two main forms:

- Hydrogen sulphide (H_2S), as a gas dissolved in the crude oil. H_2S will distil out in the lighter fractions produced by crude oil distillation, where it becomes more concentrated. It is an extremely dangerous gas, with an OEL/TLV of 10 ppm, a STEL (Short Term Exposure Limit—the maximum concentration it is safe to work in for 15 minutes) of 15 ppm and is immediately fatal at concentrations in excessive of 1000 ppm. *It has a characteristic smell of rotten eggs, and as the concentration increases it progressively inhibits a person's sense of smell. Many gassings have taken place when, after smelling H_2S, a person has been lulled into a false sense of security when the smell apparently went away, followed by collapse and death as they remained in an area of high concentration.*

BP Process Safety Booklet covering the handling of Hazardous Substances in Refineries contains details on H_2S and its effects on humans and description of incidents.

For example: A technician collapsed after removing a level transmitter from a naphtha stripper. He had started the work wearing self-contained breathing apparatus (SCBA), including checking the tower connections for blockage, but had removed it as they prepared to remove the transmitter to the workshop.

- As organic sulphur containing compounds called thiols, the most common of which are the mercaptans. These are also highly toxic and have a foul smell at low concentrations, which is why some mercaptans, e.g. ethyl mercaptan, are used as the safety stenching agent in natural gas supplies, and retail fuels such as LPG. The higher boiling point thiols can be thermally cracked to release H_2S.

Naphthenic acids

Naphthenic acids, which are cyclic carboxylic acids of the form $C_nH_{2n-1}COOH$, are present in some crude oils. While their concentration does not normally constitute an occupational exposure hazard over and above those already imposed by crude oil, they can give rise to severe corrosion within the process unit, as discussed below.

Organic chlorides

Organic chlorides can be naturally present in crude oils and can be introduced by additives injected into the produced fluids stream between the production wells and the refinery. They may not create occupational exposure problems directly, but they can initiate severe corrosion problems in downstream equipment as they can be a source of hydrochloric acid. They are neutralized by injecting amines into process streams, which is why determination of the chloride content of the crude unit overhead water stream is an important routine test to monitor chemical addition.

Salt (NaCl)

Salt (NaCl) and other water soluble inorganic compounds are normally present in the water that combines with the crude oil as it leaves the underground reservoir. Now that marine tankers have dedicated ballast tanks, the source is less likely to be sea water. While the hazards are associated with corrosion within the process unit, water associated with crude oil should always be treated as toxic and protection taken against:

- breathing in any fumes or gases released—such as H_2S;
- ingestion;
- contact with the skin and eyes.

Mercury

Mercury can be present in trace amounts in a variety of forms from elemental mercury to inorganic and organic compounds, sometimes in a complex form. Depending on concentration, mercury can present a problem as it can poison catalysts in downstream processing and can reduce product quality. Mercury may also attack some metallurgical components of downstream light ends units, and cause problems meeting effluent quality. A serious occupational exposure issue can result where mercury is deposited in the overheads condensers of crude oil and other distillation towers. It can be identified through vapour phase sampling of the column (once this is open at turnarounds) and occasionally is also evident as liquid mercury deposits.

NORM (Normally Occurring Radioactive Materials)

These are the by-products of radioactive decay within the oil bearing strata including radium and polonium radioactive isotopes that concentrate as a deposit on the internal walls of process piping and equipment. Generally this takes place at the oil and gas production facilities where it can develop into a significant problem, but can migrate to crude oil distillation processing. Where NORM is contained within the oil refinery equipment it is generally harmless although background radioactivity levels can be increased. However, it can become a significant personnel hazard through inhalation or ingestion of dust from scale formed on the inside of the equipment when that equipment is opened up for maintenance and inspection. Not all crude oils contain NORM and those that do are generally well known and precautions will be in place. However this should be considered when unfamiliar crude oil arrives for processing.

> More information on NORM can be found in the U.S. Geological Survey Fact Sheet 0142-99 'Naturally Occurring Radioactive Minerals (NORM) in Produced Water and Oil-Field Equipment—An Issue for the Energy Industry' or through in-house Occupational (Industrial) Hygiene specialists.

Solids

Solids in the form of sand and pipescale can be deposited in process equipment. The majority of solids, known as sediment, are removed in the desalter, but can still pose a hazard to maintenance and inspection personnel if deposited in downstream equipment. Appropriate respiratory protection must be worn when removing deposited solids from equipment. Laboratory tests should be carried out on crude oil imports for BS&W [Bottom Sediment and Water (i.e. sand and pipescale) which will impact the desalter performance].

Pipescale (or pyrophoric iron sulphide)

This can be a mixture of iron oxide formed during the production process of piping and steel plate that has been dislodged from the surface of the parent metal by thermal and erosive effects, and iron sulphide (FeS) that is formed by the reaction of iron oxide with H_2S in the absence of air. A major problem arises when equipment containing iron sulphide is opened up for maintenance or inspection and the iron sulphide is allowed to dry out. It then reacts exothermically with air to revert back to iron oxide, becoming sufficiently hot to create a source of ignition for a light hydrocarbon vapour cloud, or petroleum coke (discussed below). FeS is often described as Pyrophoric Scale. The process is described chemically, in simple terms:

$$2FeS + 3O_2 = 2FeO + 2SO_2 + \textit{HEAT}$$

> **WARNING: Not only is heat released when FeS oxidizes, but also the gas sulphur dioxide, SO_2, *which is toxic*. However, SO_2 does not inhibit the sense of smell in the same way that H_2S does; instead it creates a strong choking sensation that forces affected people to leave the area. The key to preventing these so-called pyrophoric fires is to keep the internals and equipment wet when exposed to the atmosphere.**

Petroleum coke

Petroleum coke is produced through the thermal cracking of hydrocarbons within the high temperature sections of the plant. Provided normal operating conditions are not exceeded the laydown of coke will progress at a low or zero rate over the production run of the process. However, where there are significant process disturbances, coke production may occur more rapidly, leading to blockage or damage to fired heater tubes and other column internals. These effects are discussed more fully later in this booklet. When removing coke personnel should be protected from inhaling dust and from contact with the skin, where it can be an irritant. Longer exposures may cause dermatitis or even have carcinogenic properties.

> **WARNING: Coke deposits should be kept wet when exposed to air as they can easily be ignited by pyrophoric scale formed in the same location. Clearly, depending on the quantity of coke available, this could result in a very destructive fire within vessels and columns. Wetting will also reduce the inhalation hazard.**

Coke and pyrophoric iron sulphide tend to migrate into column packing. In practice it is difficult to ensure that all parts of a column packing are properly cleaned and continuously water-wetted during shutdown. There are chemical cleaning techniques that can be used to remove coke, but these do not give any guarantee of effectiveness, particularly if a column packing is badly fouled. Structured packing in crude distillation equipment is generally constructed from stainless steel. If other materials, such as titanium and aluminium, are used to make the packing, a separate hazard analysis would be required to address specific combustion properties of these materials.

The photograph below shows the result of a pyrophoric fire involving petroleum coke deposited within a packed section inside a VDU main column. This occurred during a turnaround. The outside diameter of the column bulged by 160 mm (6 inches) due to the fire. If the coke deposits and iron sulphide scale had been kept thoroughly wet before and while they were being removed this damage would not have occurred [see also section 3.3 on page 32].

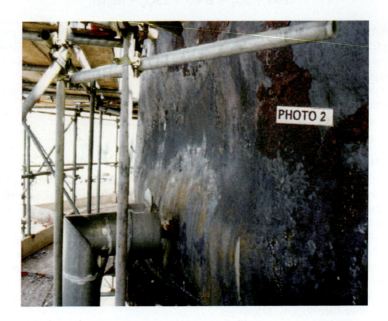

2.3   Other hazardous substances in common use on CDU/VDUs

Superheated steam

Superheated steam is used to strip light ends from the side stream distillates and residues on both CDUs and VDUs. It is usually taken from the LP (low pressure) steam header which operates at between 3 and 5 barg (45–60 psig) and is in turn typically supplied through letdown from the MP (medium pressure) 10 barg (150 psig) steam supply through small steam turbines or pressure reducing stations. It is often superheated in the convection bank of the main charge heater to a temperature of about 300°C (572°F).

BP Process Safety Booklet *Hazards of Steam* contains information on the properties and hazards of steam.

WARNING: Superheated steam is invisible at the source of a leak, only becoming visible when droplets start to form in the cloud, way away from the point of release.

Water

BP Process Safety Booklet *Hazards of Water* gives information on the properties and hazards of water.

Note particularly the section on water in vacuum columns.

- **Water is used in the desalting process** injected upstream of the desalter at the inlet to the pre-heat exchangers and/or immediately upstream of the desalter itself. It can be sourced from other refinery processes, such as Sour Water Strippers (SWS), designed to remove dissolved H_2S and ammonia in aqueous effluents from FCCUs, hydrocrackers and hydrotreaters. However, if the SWS is not performing as designed, for whatever reason, there could well be a breakthrough of dissolved hazardous gases. The SWS effluent also contains dissolved inorganic salts, which can be hazardous through contact with the eyes, by inhalation of mist, ingestion or through skin absorption.

- **Water leaving the desalter** contains greater concentrations of dissolved inorganic salts than the water injected upstream. It can also contain hydrocarbons if the desalter interface level control is not working correctly or through a lack of de-emulsifying chemical addition. In addition to the hazards described above, desalter effluent is hot and when hydrocarbons are present can, when released to atmosphere, create clouds of flammable and toxic vapour.

WARNING: Particular care should be taken when proving the location and depth of the desalter hydrocarbon/water interface layer by using the tri-cocks provided for this purpose as they discharge to atmosphere, albeit through small bore connections. There can still be sufficient release of hazardous materials to overcome operators.

- **Water accumulating in the CDU/VDU overheads accumulators/reflux drums** is the combination of stripping steam that has condensed in the cooler parts of the main columns and any water carry through from the desalter. It contains dissolved inorganic chlorides as well as ammonium salts. This stream is regularly sampled for pH by operators who should ensure that they are wearing eye and face protection and impervious gloves.

WARNING: Water from overhead accumulators and reflux drums can contain dissolved H_2S.

Ammonia

Ammonia is occasionally injected on some older CDUs to neutralize inorganic acids in the CDU and VDU overheads. A better and safer practice is to use neutralising amines rather than ammonia.

Particular care must be taken when handling anhydrous ammonia as the OEL/TLV is 25 ppm, with an STEL of 35 ppm. It is flammable, although difficult to ignite in the open, is toxic by inhalation, can cause burns to exposed skin and is toxic to aquatic organisms, which means that any spillage must be diluted with copious amounts of water.

> **WARNING: Respiratory protection in the form of positive pressure air line or SCBA will be required when dealing with ammonia spillages.**

Ammonia is usually delivered in transport containers that are either connected to the process directly, or discharged into a storage vessel on the unit. In addition to PPE requirements, it is good practice to install a safety shower alarmed back to a permanently manned control room adjacent to where connections are made and broken.

> BP Process Safety Booklet *Hazardous Substances in Refineries* contains information on ammonia and its hazards, together with some accident case histories.

Chemicals used for breaking desalter emulsions and for corrosion control

There are a large number of chemicals used for the above purpose, provided by a number of suppliers. They all have varying degrees of flammability and toxicity, and some may be static accumulators. Some may be supplied in IBCs (Intermediate Bulk Containers) that are transported by fork lift trucks, and others in steel drums.

> **Operators must ensure that they are fully aware of the specific MSDS (Material Safety Data Sheet) supplied by the chemical manufacturer, be fully trained in the handling of the chemicals and their containers, and understand the PPE and any respiratory protection that is required to be worn. Forklift trucks must be approved for use in the hazardous area they are required to work in to deliver the chemical containers and remove empty ones.**

Nitrogen

Nitrogen is used for purging equipment to prepare it for work to carried out at shutdowns/turnarounds. Permanent supplies of nitrogen may be required to purge some instruments, particularly analysers.

> BP Process Safety Booklet *Hazards of Nitrogen and Catalyst Handling* contains information on the properties of nitrogen and a large number of accident case histories.

WARNING: Where nitrogen is used or released in a confined environment, such as an analyser house, access to the inside of that enclosure must be strictly controlled.

Nucleonic sources

Nucleonic sources are used to measure liquid levels in some process columns and vessels, normally where the level is difficult to measure with conventional displacer-type level detectors. The source and detector are located on opposite sides of the column or vessel. The source emits gamma rays that are focused into a narrow beam that passes through the column or vessel walls and the process fluids contained inside. These are sealed sources that are encapsulated by the manufacturer and provided with a window that can be opened when the device is installed in its operating position. The arrangement is very safe when installed correctly. When the process equipment is opened for internal inspection and cleaning, the window on the source container *must be locked in the closed position*, or the source removed in its container with the window locked in the closed position for secure offsite storage.

The locking open and shutting of the window must be controlled, with the involvement of the site Radiation Safety Officer, under a lock out/tag out procedure in the same way as other isolations are made safe.

WARNING: Be alert for the radiation warning signs around base level measurement devices on CDU and VDU main column bottoms.

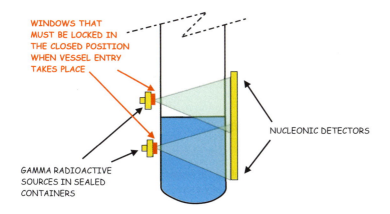

WINDOWS THAT MUST BE LOCKED IN THE CLOSED POSITION WHEN VESSEL ENTRY TAKES PLACE

NUCLEONIC DETECTORS

GAMMA RADIOACTIVE SOURCES IN SEALED CONTAINERS

3

Physical hazards

An analysis of the failure mechanisms associated with over 100 CDU/VDU incidents showed that the largest failure mechanism was associated with human error, followed by corrosion, followed by unforeseen process upsets of varying kinds. These are discussed below, with the exception of corrosion which is discussed in Chapter 4.

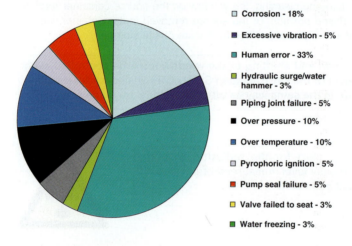

- ☐ Corrosion - 18%
- ■ Excessive vibration - 5%
- ■ Human error - 33%
- ■ Hydraulic surge/water hammer - 3%
- ▨ Piping joint failure - 5%
- ■ Over pressure - 10%
- ■ Over temperature - 10%
- ☐ Pyrophoric ignition - 5%
- ■ Pump seal failure - 5%
- ☐ Valve failed to seat - 3%
- ■ Water freezing - 3%

The survey of CDU/VDU accidents that have taken place in a major oil company's refineries since 1971 has provided the summaries that are contained in the paragraphs below that are highlighted in yellow (numbers in brackets refer to incident list in Chapter 9).

3.1 Start-up and shutdown

Human failure

- When starting up any process unit, great care must be taken to ensure that venting and draining that is not normally associated with normal operation is carried out safely. Piping and vessel drains will trap debris and the products of corrosion, much of which can become dislodged and migrate to the lowest points in the process system as temperatures and flowrates are increased during start-up. Blockages such as this often become consolidated into a solid plug that will prevent any release of process materials until the valve has been opened a considerable amount. The sudden high rate of discharge that then results coupled with splashing and gas release can cause the operator to retreat, often injured, with the valve being left open. Access to the release point then becomes impossible with flammable liquid and vapours travelling considerable distances before ignition occurs.

Vapours were ignited after being released from a manually opened 2″ atmospheric drain valve on the CDU overheads condensers, which had suddenly cleared. The operator, who received minor burns, managed to close the valve and stop the release. (138)

During a CDU start-up, liquid and vapour escaped and ignited when an operator drained a main column overheads condenser to atmosphere. The 2″ drain valve was blocked with scale and suddenly cleared as the valve was progressively opened. (22)

- Process units are large and complex, with many potential sources of release unless great care is taken to ensure that all flanges, piping caps and equipment plugs are correctly installed and tightened prior to start-up.

The vent plug on a CDU main column base level gauge glass was left finger tight when the glass had been last used to check a high level. A second operator then used the glass to check the column level, whereupon the plug was ejected and hot oil released. (110)

A flanged joint on the CDU overheads system failed due to flange bolts being insufficiently tightened during the preceding turnaround. (13)

- In some cases, vents or drains that had been opened during the previous shutdown had not been fully closed prior to start-up.

A fire occurred when naphtha was released from an atmospheric vent line on the VDU reflux drum during plant recommissioning following an emergency shutdown earlier in the day. (82)

A flange had been cracked open on the desalter pressure relief line to facilitate steaming after shutdown. Oil leakage onto the piping escaped and was ignited by hot equipment. (17)

- Start-up of fired heaters is where the largest number of accidents occur during start-up. All the following occurred due to some form of human error in not following proper procedures or good practice.

Two explosions occurred within a 3 cell CDU charge heater when relighting after a power failure. The fireboxes had been purged using the forced draught fan, but the fan louver was stuck in the closed position and the low pressure alarm was disconnected. (112)

A severe internal explosion occurred in the radiant section of a CDU charge heater while it was being recommissioned after turnaround. Gas had entered the firebox from the process side through tube header plugs that had not been reinstated. (30)

An internal explosion occurred in the VDU charge heater firebox during start-up. Delayed ignition occurred due to the high flow of waste gas into the firebox through ducts in the floor coincident with low fuel gas firing and high process temperatures. (114)

An explosion occurred in the CDU charge heater firebox and ducting during relighting of the pilot burners after an emergency shutdown. Fuel gas isolation procedures had not been followed as relighting was being done in haste to avoid shutdown of the Fluid Catalytic Cracking Unit (FCCU). (44)

An explosion occurred in the convection section of the VDU charge heater during lighting up. The blind in the main fuel gas supply had been removed prior to attempts being made to light pilot burners. (32)

- It is usually a feature of CDU and VDU charge heaters and main columns that there are no block valves fitted in the transfer line. This is because under normal conditions there is no requirement to decoke the heaters outside of the normal turnaround interval. At turnaround, the two are blinded from each other to allow different activities to be carried out concurrently. A major problem can occur when attempting to remove the blind (spade), particularly if one side is further forward in the re-commissioning programme. An example is where the heater activity is in-situ decoking using a mix of steam and air to burn the coke from the inside of the heater tubes, venting to atmosphere immediately upstream of the heater/main column blind. As the heater main burners are required for the decoke, the correct procedure should have been to shut the heater down completely and isolate the main and pilot burner fuel supplies. However, in the examples below, in order to save time, pilot burners were left in commission while the blind was being turned—with fatal consequences in one case.

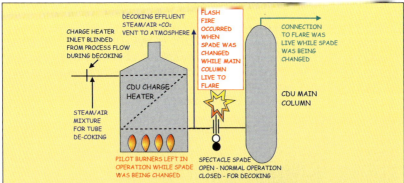

An explosion occurred during the removal of a 24″ blind from the transfer line between charge heater and main column after decoking, killing one man andinjuring six others. The column was not gas free. Escaping hydrocarbons were ignited by pilot burners. (41)

On another incident, the blind on the transfer line outlet of a CDU charge heater was being swung to the open position after decoking. The column side of the blind was open to the refinery flare system, releasing gas that was ignited by pilot burners. Fortunately only minor injuries occurred to one worker. (38)

- Checking of the condition of plant and equipment. Plant and equipment that has been shutdown, some with components repaired or replaced should be regularly inspected to identify any problem areas before they lead to a major accident.

The shaft of a vacuum residue pump circulating gas oil was seen to be running red hot, most probably because of a bearing failure. The pump seal failed before remedial action could be taken. Gas oil escaped and ignited (103).

It should be noted that some streams on a CDU and VDU are operating above their auto-ignition temperature. So in the event of a seal failure, it is highly likely that the material will auto-ignite immediately.

- Draining of water from process equipment prior to start-up.

The hazard of water entering hot equipment is discussed throughout this booklet and occurs most frequently at start-up. The example below describes the consequences of inadequate water draining.

During start-up of a CDU, the bottom pumparound pump was commissioned. Shortly afterwards the CDU main column atmospheric relief valve (ARVs) lifted for three minutes. 30% of the release fell outside the refinery fence onto cars and homes. Many column trays were destroyed. A 3 metre (10 ft) length of pipe above a check valve on the pump discharge had not been drained of water, which resulted in about 300 litres (80 US gallons) of water being introduced into the hot column.

Integrity of piping and equipment

- Quality assurance and control procedures must be rigorously enforced during the construction and maintenance processes. Where mistakes have been made they only become apparent at start-up. See also section 4.1.

A carbon steel coupling connecting a thermowell to a VDU transfer line was found to be leaking when insulation was removed during cold circulation prior to start-up. The transfer line and thermowell were constructed of the correct alloy steel but the coupling was not. (2)

A leak was observed on the VDU transfer line during cold circulation prior to start-up. A carbon steel pipe had been welded to the alloy steel transfer line and alloy steel thermowell. A contractor had used non-standard materials and construction methods. (93)

- Hidden areas of exceptional vulnerability to failure.

Some areas are prone to advanced corrosion and if not identified and removed at shutdown, can become a major issue at start-up. The most common of these are 'dead legs' created in piping systems when changes are made to the original design. While removal of redundant pipework is necessary to avoid these areas being created, it undoubtedly adds to project cost and schedule, leading to decisions to blind off redundant sections of line with an intention to make good at a future opportunity, which may be subsequently forgotten. See also section 4.1.

> During plant start-up leakage was observed at an uninsulated flange to valve connection. Severe high temperature sulphidation corrosion had occurred in a dead leg within which there had been a thermosyphon effect that had kept the piping hot. (119)

Unstable process conditions

- Start-up and shutdown are periods of extreme mechanical and thermal stress on piping and process equipment. Even small mistakes can take on a much greater significance.

> A VDU suffered severe mechanical damage to internals during start-up due to excessive water lying in the system, most likely admitted through a leaking heater pass stripping steam connection. High charge heater temperatures caused coking in tubes. (50)

> During start-up of the CDU/VDU after plant precautionary shutdown in anticipation of a hurricane, a flanged joint on a valve at the end of a 'dead leg' piping section failed due to severe hydraulic surge created during the preceding shutdown. (70)

Fitting of drain plugs on valved drains and vents

It is considered good practice to fit plugs to valved drains used at start-up to drain down equipment for maintenance while a unit is on stream. On the face of it a minor housekeeping mistake, but one that can have major consequences as valves can work open by vibration alone. In the example below it is highly likely that the drain connection would have accumulated debris over time which allowed the valve to open significantly before the blockage cleared, creating an unstoppable situation.

> Atmospheric residue at 379°C (625°F) was released from a 19 mm (3/4 inch) drain valve on the body of a pressure control valve. The drain had not been fitted with a steel plug. The drain valve was vibrated open when a relief valve on the residue line lifted. (31)

Resorting to unsafe means to resolve a problem

- Access to overhead valves and equipment is not always provided for start-up and shutdown. In this case an operator tried to gain access without proper facilities, with tragic results.

> An operator fell to his death from an overhead piperack while attempting to open two 1″ valves at a steam header. There was no permanent or temporary access platform available, and no procedure had been followed. (40)

- Sometimes, a problem arises that needs rapid resolution to keep a start-up on schedule. Unfortunately, in many cases, unconventional remedies result in serious consequences.

21

CDU distillate hydrotreater charge pumps with a shut-in head of 75 barg (1100 psig) were used to slop light product from the main column. A low pressure vessel in the Class 150 tankage area ruptured when an offsite valve was closed against these pumps. (33)

Equipment deteriorated to a poor condition after a long period of operation

When process units are started up after a major turnaround, the condition of the equipment is usually returned to a state of integrity similar to the design. Clearly, after a long period of operation, equipment may have deteriorated to an extent where all corrosion allowances and similar aspects of plant and equipment designed to wear out have been taken up. It is, therefore, necessary to treat plant and equipment with extra care as it is shutting down depending on the observed state of deterioration.

Process tubes within the charge heater that had been severely thinned due to internal coking and high heat flux failed when water hammer was created whilst steaming through to the column. (5)

Preparing equipment for turnaround activity

As systems become available during a unit shutdown, operators start to prepare these for the subsequent turnaround. Even more risk assessment, care and attention needs to be taken than for any maintenance task that is carried out with the unit on stream. With an impending turnaround the opportunity may be taken to carry out more simultaneous works, and the installation of temporary access facilities, such as scaffolding may hinder access to and from the equipment being prepared.

A technician was opening a flange on an incoming naphtha line that had not been drained down. A release occurred which was ignited by adjacent steam pipes. (29)

Spent caustic contaminated with hydrocarbon was flushed into a wash drum that had been cleaned and steamed in preparation for a turnaround. The drum was allowed to overflow with escaping gasoline being ignited by a steam line. (39)

Problems with dismantling process equipment and piping

Experience shows that on many occasions equipment cannot be dismantled in accordance with maintenance procedures due to accumulation of deposits, damage to internals, etc. When this happens it is essential that time is taken to review the situation and to prepare a new risk assessment and safe work procedures to try to resolve the situation. These would include checking equipment after steaming, for example, to ensure that it has been thoroughly depressured before attempting to proceed with dismantling.

Steam at 8.6 barg (125 psig) had been injected overnight into the shell of a heat exchanger when attempts to remove the cover had previously failed. The following day the cover was violently released killing one man and severely injuring three others. (8)

Unacceptable status of process unit condition after an unscheduled or emergency shutdown

- There is a history throughout the industry of going to extreme lengths to get plant back on stream after an unscheduled outage. This carries a high risk, especially if start-up checks are rushed in an attempt to speed up the process. There is no substitute for ensuring that a plant is in a thoroughly safe and stable state, with all the checks and balances made, before attempting to remedy faults and restart the unit. This is called a *Pre-Start-up Safety Review (PSSR)*.

A CDU charge heater had been completely shutdown after a major process upset resulted in liquid entering the fuel gas system. Individual gas burner isolation valves were not closed. An explosion occurred when attempting to drain the fuel gas system. (113)

- In some cases the cause of the unscheduled shutdown will have damaged plant and equipment. Extreme care needs to be taken to ensure that when investigating the status of the equipment, every effort is made to ensure that this can be done safely. In the example below two faults contributed to the outcome, but each was a hazard in its own right—contaminated nitrogen (which would have had flammable and asphyxiating hazards of its own that need to be taken into account in a risk assessment) and pressure that was retained inside the damaged heater tube.

After a shutdown caused by a fire on a CDU charge heater, it had been steamed and nitrogen purged. When checks were made at heater tube plugs for leakage, one became dislodged with escaping hydrocarbons from contaminated nitrogen igniting. (25)

Equipment damaged by thermal cycling due to repeated start-up and shutdowns

The repeated shutting down and restarting of a process unit or piece of equipment can have a damaging effect on its integrity.

Flange bolts loosened by thermal cycling through successive start-up/shutdowns released heavy gas oil at a draw off connection on the main column. This connection had been previously blanked off as it was no longer required. (45)

3.2 Normal operation

Many of the incidents reported are equipment specific, and these are discussed in Chapter 4. Others were shown to be due to some form of human error.

Human error

• Failure to open or close valves that have been opened temporarily, or vice versa, has led to a number of major accidents. In some cases, such as the residue pump drain described below, the fact that a valve has been left open is not immediately apparent if, for example, it has been plugged with solidified heavy oil which subsequently melted.

Liquid hydrocarbon entered the waste gas stream routed to the VDU charge heater when a bypass around the liquid knockout facility was left open. An external fire occurred that was quickly extinguished when an operator correctly diagnosed the problem. (91)

The drain line of a CDU residue pump was open but plugged with solidified residue when it was handed back to operations from maintenance. The plug melted as the pump was being recommissioned, spraying the surrounding equipment with 330°C (625°F) residue. (95)

Heavy oil wax and scale that had prevented the drain valve on a VDU overflash pump drain from being fully closed at a previous turnaround, melted when the pump was recommissioned. VDU overflash at 315°C (600°F) escaped and auto-ignited. (129)

• Sampling of hot oils can be particularly problematic. The correct procedure using the right equipment must always be used, but in many cases when shortcuts are taken it can result in a major incident occurring.

During sampling of hot vacuum residue, sample valves were cracked open. A sudden uncontrolled release at 300°C (570°F) occurred which auto-ignited. A sample cooler was available, but it had been decided not to use it despite the temperature of the residue being greater than the autoignition temperature. (127)

A drain valve on a VDU bottoms system had been identified as a sample point for a test run but no flow occurred when it was opened. The valve was left open. The blockage suddenly cleared creating a major release which ignited immediately. (123).

• Making rapid or large incremental changes to operating conditions. Changes made to operating conditions on process units should always be carried out smoothly and within prescribed limits, e.g. increase throughput in 2,000 BPD (15 m³/h) steps; increase temperatures at no more than 25°C (45°F) per hour.

An operational upset at the CDU/VDU resulted in the release of vacuum residue from the top of a blowdown stack. Some heavy oil spray went offsite. A large increase in crude feedrate had occurred after change over of the desalted crude pumps. (60)

Change over of desalted crude pumps led to a large increase in feedrate to the CDU resulting in light material feeding into the VDU. This caused a loss of vacuum and release of about 60 barrels (10 m³) of heavy oil from a pressure relief system to atmosphere. (87)

Maintaining close surveillance of operating equipment

One of the essential roles of the outside operator is to maintain close surveillance of operating equipment at all times, including when operating conditions are steady. One frequent cause of failure is where screwed fittings (which should be used only for final connection of small bore equipment and non corrosive service) become detached through vibration. In many cases this can be predicted where large objects, such as pressure gauges, are supported purely by a low schedule section of piping.

A pressure gauge worked loose and came off the inlet piping to a VDU fired heater. Hot oil at 316°C (600°F) ignited immediately. The gauge had been installed incorrectly twelve months previously at a turnaround. (89)

A pump casing drain piping suffered from severe high temperature sulphidation corrosion that resulted in the threaded section (see red arrow) of the drain pipe engaged in the pump casing thinning to a point where the threads separated under the load from the pressure within the pump. The drain piping was blown out of the casing, and distillate fluid released at approximately 360°C (680°F) auto-ignited. (139)

Practising good housekeeping

Good housekeeping is another essential task of the outside operator. When this is ignored or missed because of more urgent work to keep a process unit on stream, the consequences can be a major incident.

A flexible hose, that had previously been used to transfer corrosion inhibitor dissolved in a naphtha solvent, was used to wash down a process unit. Flammable material was sprayed over hot equipment and it ignited. The fire was made worse by poor housekeeping. (101)

Process systems becoming overpressurized

- A simple failure of a single process controller can have major implications, resulting in process systems becoming overpressurized

> A serious process upset resulting from failure of a level controller caused pressure to rise in the CDU pre-fractionator, main and vacuum columns, approaching the set pressures of the atmospheric discharge pressure relief valves. An environmental incident may have occurred if the relief valves had opened. (74)

- Another overpressurization situation can result from large quantities of water entering the process units. It is for this reason that most sites stipulate a 48-hour minimum settling time for crude oil tanks that have been filled from a marine tanker. Also, good desalter operation can mitigate water slugs, but in some cases intermediate feeds can be processed direct to units effectively bypassing the desalter.

> Substantial quantities of water entering the VDU feed drum during a feedstock change caused the RV to lift discharging a large volume of vapour into the main column, resulting in leakage at a transfer line flange. The resulting fire spread quickly. (71)

- Pressure can also build up in static situations, such as 'boxed in' piping and valves, particularly in close proximity to hot equipment.

> Gas oil leakage from the bonnet joint of a valve only used during CDU maintenance leaked onto a 300°C (570°F) hot pump and was ignited. (130)

- Another overpressurization situation can occur where the warm-up bypass of a hot oil pump is commissioned before that pump's suction valve has been opened. Almost all pumps are designed with the suction flanges and pipework, including strainers, at a lower flange rating than the discharge. The warm-up bypass allows hot material from the running pump discharge to enter the body of the standby pump, which then flows back through its suction line. See also section 4.4.

> Hot oil escaped from a vacuum residue pump strainer when its associated pump was blocked in with the warm-up bypass open to the adjacent running pump, subjecting the suction filter to discharge pressure. (9)

Low temperature—water freezing

Many refineries operate in areas of the world where winter weather can produce freezing temperatures over a considerable time. Where water has been allowed to remain in systems where there is little or no flow, freezing will take place, with the potential for a major release when the systems subsequently thaw out. This can occur even on hot equipment, particularly when drains are uninsulated and exposed. Particular care must be taken to drain down systems where water can accumulate when those systems are shutdown.

Problems with the level detector on the CDU kerosene side stream stripper had caused the equipment to be drained frequently during cold weather. Hot kerosene escaped and ignited when a drain valve was blocked by a plug of ice that subsequently thawed. (92)

Water trapped in the gasket area of a flushing oil line to the VDU column bottoms froze and ruptured the gasket. Accumulation of water was due to intermittent use of the flushing oil. Escaping flushing oil was ignited. (78)

Water collecting in the body of a ball valve located in a 3″ Class 300 crossover line to the standby stabilizer pump froze, resulting in the failure of the valve body radial flange bolts. The escaping butane/propane was ignited at a nearby fired heater. (23)

Water freezing within a 2″ carbon steel pipe caused it to fail releasing a high pressure spray of naphtha towards a fired heater, where it ignited. The failed pipe, which formed a 'dead leg', had not been used in 20 years, and was not fully isolated. (42)

A 3″ steam stripping line that was shutdown accumulated aqueous condensate which froze in cold weather, rupturing the pipe. Naphtha escaping from the rupture ignited. An open block valve and failed Non-Return Valve (NRV) prevented isolation from the crude tower. (77)

Magnitude of a fire following pipe rupture on a CDU naphtha stripping steam line that had been isolated and accumulated condensate. The condensate froze during cold weather causing the pipe to fail when it subsequently thawed.

Close-up view of rupture showing maximum deformation region (ductile tearing) at red arrow.

Thermal shock

- Process units that experience thermal shock can quickly lose containment with a major fire potential. This is because when systems are heated up in accordance with operating procedures, e.g. no more than 25°C (45°F) per hour, piping flanges and bolts heat up together maintaining their integrity. Sudden changes in temperature upset that equilibrium resulting in loss of containment. Where such a release ignites, the fire quickly intensifies as the bolts expand far more quickly than the flanges.

> Following temporary loss of feed to the VDU, leakage of atmospheric residue from thermally shocked VDU feed/overheads heat exchangers sprayed over adjacent equipment and ignited. (28)
>
> As a CDU feedrate was being reduced due to unit instability, a charge heater pass control valve closed. The valve was reopened on manual but a small fire had already started and the unit was shutdown. (24)

- The thermal equilibrium that maintains tight flanges can be upset when these flanges are insulated after the design conditions have been reached.

> A valve fitted between flanges using long bolts was insulated for energy conservation while the valve and pipework were at the normal operating temperature of 310°C (590°F). Some hours later, the flange opened when the long bolts expanded with the heat—a major release occurred which immediately auto-ignited. (107)

- In addition to thermal shock, thermal cycling, i.e. repeated heating and cooling albeit within safe operating limits, can result in loss of containment.

> Failure of the CDU main column recirculating pump's mechanical seal assembly allowed hot oil to escape and auto-ignite. The pump had frequently been started up and shutdown causing seal assembly bolts to loosen over time through thermal cycling. (68)

Power failure

- Impact on fired heaters. Most process units are designed for loss of major utilities, for example, with fired heaters, having the main and pilot burner fuel supplies automatically shut off. However, where an off-gas stream is burned in a heater because it is at too low a pressure to be routed to flare, the manual shutdown of this flow to the heater is sometimes forgotten.

> A ground fault and too high a local breaker setting resulted in a total loss of electrical power to the CDU/VDU when operators attempted to change over the vacuum bottoms pump. This electrical fault caused a VDU shutdown. Vacuum tower off-gas continued to be fed to the charge heater. While flow through VDU passes was maintained by steam turbine driven pumps, flow through a naphtha convection bank fed by electrically driven pumps stopped, resulting in a major fire when the tubes failed. (60 & 61)

- Impact from lifting atmospheric pressure relief valves (ARVs). Where fitted, ARVs are generally sized for total power failure to the process unit. When they lift they can create a hazard. That is why many ARVs have been replaced with relief systems discharging to closed flare systems—but this is a major modification. See also section 6.4.

Atmospheric Relief Valves lifting on a CDU main tower

Impact of operations on an adjacent unit

- Process units do not operate in isolation. Process upsets on neighbouring units can have a major impact on each other, particularly where they are thermally or mechanically integrated.

A major upset on the FCCU caused unburned hydrocarbons from the regenerator entering the thermally integrated firebox of the CDU resulting in tube failure through excessive tube skin temperatures [810°C (1400°F)]. (36)

Liquid hydrocarbon carryover into the sulphur plant incinerator resulted in a fire in the common stack with the CDU. Initial upset in the CDU desalter escalated to pre-flash column and naphtha stabilizer with liquid hydrocarbon carryover into amine system. (57)

- Alternatively, the impact can be purely through close proximity.

A fire of 20 minutes duration occurred beneath the VDU fired heater as a consequence of froth-over at the adjacent Bitumen (Asphalt) Unit during its start-up. (3)

Maintaining the integrity of standby equipment

While the focus is on maintaining the running equipment in good order, there must be an equal duty of care to standby equipment that may auto-start or have to be commissioned quickly.

A small fire occurred at a CDU secondary feed pump when the standby pump auto-started as one of the two duty pumps tripped (this was a 2 out of 3 arrangement). Mechanical seal leakage was attributed to bearing failure through lack of lubrication leading to misalignment of the shaft. (99)

Use of temporary equipment

Equipment that is included in the design of a process unit has been selected to make it compatible with the process conditions and hazardous area classification. When temporary equipment is needed because the installed equipment has failed, for example, great care must be taken to ensure that it is fit for service and the location where it will be used.

Released hydrocarbon from a pipe clamp connection was ignited by a diesel engine driving a temporary hydraulic pump used to flush coils of the residue cooler box. This pump replaced the CDUs steam reciprocating pump normally used for this duty. (52)

High pressure water jetting with the unit on-stream

High pressure water jetting is an essential task when shell and tube heat exchangers foul during normal operation. It can be carried out safely provided the preparation and isolation of the work site is carried out correctly, supported by a thorough risk assessment. Invariably, the work is carried out by contractors who may not be aware of the high standards required in training and supervision. Careful attention needs to be paid to contractor selection.

A technician was seriously injured while cleaning a fouled heat exchanger with a high pressure water jet. He had stumbled when the jet was suddenly shut off, after which the jet had recommenced as the tip of the lance contacted his right thigh. (106)

During cleaning of a heat exchanger with a high pressure water jet, a worker suffered a broken arm when a hose connection came apart and the unrestrained hose lashed around. The threaded connection had not been screwed on tightly enough. (116)

A high pressure water release that impacted the operator occurred during start-up of a high pressure jetting pump when the overpressure bursting disc failed or lifted. Two of the three jetting nozzles were found to be blocked by grit. (121)

3.3   Unit turnarounds

Harmful deposits created when the process unit was in commission

There are a number of examples of where deposits created in normal operation of a process unit remain benign until the unit is opened up for inspection and maintenance. The example below appears to indicate that during operation, the draught controls on the heater were sufficiently relaxed to create positive pressures within the firebox and convection sections.

Hygroscopic combustion deposits entered the header boxes of a CDU charge heater during normal operation. Severe corrosion of the tube headers occurred when these deposits hydrated during shutdown leading to failure when the unit recommissioned. (109)

Another example is the formation of polythionic acids inside stainless steel process equipment and systems when they are opened up to atmospheric humidity, leading to stress corrosion cracking (API RP 571).

Hazards when handling heavy components when they are being dismantled from equipment that has previously been pressurized

When a process unit has been shutdown and prepared for turnaround, there is still a danger that equipment may not have been fully depressured through release of purging media, such as nitrogen. In some cases inerting of equipment by maintaining a slight positive pressure of nitrogen may have been done deliberately to keep out air and moisture. Even small system pressures can create large forces when they act over even a moderate area.

A fired heater tube plug weighing 9 kg (20 lb) was ejected from the header box by internal pressure, travelling a distance of 30 metres (100 ft). The heater tubes had not been depressured, and the work permits were inadequately prepared. No ignition occurred. (104)

Contamination of essential utilities

Operations and maintenance functions have come to rely on the inert properties of nitrogen to ensure that systems are maintained in a safe state. However, this reliance breaks down when the inerting agent becomes contaminated with air/oxygen or flammable materials. Strict segregation and isolation requirements must be met at all times.

A leak found during a final pressure test was being repaired under nitrogen cover, but a fire kept recurring. It was found that the nitrogen supply was contaminated with hydrogen as the two gas systems were permanently connected with only valve isolation. (131)

The hazards of pyrophoric materials and deposited coke

As described previously, two materials created inside process equipment, particularly at high temperatures and with sulphur containing hydrocarbon, can combine to create a situation that can result in major damage to process equipment when it is opened up to the air.

Refer to the BP Process Safety booklet *Safe Ups and Downs for Process Units* for more guidance on this topic.

Pyrophoric scale/coke ignited when air was admitted to the VDU column. A major internal fire overheated the shell of the main column causing it to buckle. (58)

Air entered the VDU tower overheads system causing an explosion ignited by pyrophoric deposits. There were signs of burning in the light gas oil structured packing beds and distortion of one of the heavy gas oil pumparound trays below. (64)

When the CDU main column was opened up after a shutdown caused by a major leak, an internal fire was ignited by pyrophoric iron sulphide that 'collapsed' the column. (120)

See also section 2.2.

Isolation of large items of process equipment

With the increased size of some process equipment, such as CDU and VDU main columns, the resources to effect positive isolation by blinding at very large nozzles, such as those at the overheads connections, are immense, leading to a serious challenge to leave the systems connected and treat the combined systems as one for the purposes of Control of Work. However, the downside of this is that severe limitations must then be imposed on works on the resultant common system that are carried out concurrently, adding to turnaround time and programme complexity. When the systems of control break down, the results can be fatal.

Two operators were asphyxiated while inspecting the top section of the main column. Hot work created a fire within the overheads line at the relief valve header, which was not isolated from the column. Hot gases were swept into the column by the chimney effect. (48) *This incident is described in detail in section 6.1* (see page 60).

Hazards of de-isolating large systems

When unscheduled shutdowns occur on process plant containing large systems, every effort is expended to get repair work completed and the unit recommissioned. Occasionally, particularly where large systems are involved, there may be resistance to isolate to a turnaround standard, as shown by the example below. See also section 3.1.

During removal of the 42″ blind between the CDU main column and overhead line a flash fire occurred. The tower and overhead line had been previously purged with nitrogen to repair a leak on the overhead line. Fuel gas had entered the column from the flare header and overhead drum while the repair work was carried out. This was ignited by temporary halogen floodlights. (69)

Hazards associated with post turnaround pressure testing

- Extreme care must be taken to differentiate between system pressure testing to confirm the mechanical strength of the system components and leak testing to ensure that there will not be loss of containment when process fluids are introduced into the system. In the case of pressure testing metallic components, the pressure vessel and piping codes can call for up to 1½ times the design pressure to be applied to the system. For leak testing it is only necessary to achieve the normal operating pressure. Some process equipment components are not capable of being tested to pressure test levels, although being fully capable of handling normal operating pressures.

During a hydraulic test on a crude distillation unit desalter vessel, a porcelain insulator shattered. Fragments were ejected over a wide area through an open flange on the insulator duct but there was no injury to personnel.

It was concluded that the insulator was cracked at the clamping point when the securing bolts were tightened and had disintegrated under the test pressure.

To avoid a recurrence, the insulator gasket landing area was increased and the bolt tensioning monitored during installation. (16)

- After piping and equipment modifications carried out mainly during a turnaround, the mechanical strength of the process system needs to be confirmed by pressure testing. Usually, pressure testing carried out during the original construction is by hydrostatic testing, i.e. filling the entire system with water until it is hydraulically full and then applying a pressure using some form of specially constructed pump. This is considered to be a safe method of testing as the stored pressure is immediately released should a failure occur. However, in some cases after modifications to existing plants, there may be a need to pneumatically test the system in order to avoid getting it wet with the subsequent problems of draining down and drying out. Pneumatic pressure testing is far more hazardous than hydraulic testing (due to the large amount of stored energy in an expanding gas stream) and should be avoided as far as possible as it may lead to potentially serious safety incidents.

4

Hazards relating to equipment failure

This section discusses specific equipment hazards. In addition to corporate experience, input has been obtained from API RP 571 *Damage Mechanisms Affecting Fixed Equipment in the Refining Industry*. Process engineers and operators should alert corrosion engineers and plant (metal) inspectors allocated to their process units when safety critical wash water or chemical injection streams do not function as required in plant operating procedures or instructions.

An analysis of an oil company refinery major accident reports for CDU/VDUs yielded the following analysis:

Equipment failures—% incidents reported.

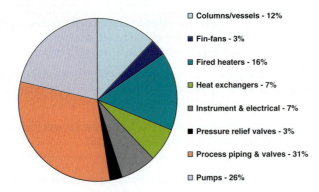

- ☐ Columns/vessels - 12%
- ■ Fin-fans - 3%
- ■ Fired heaters - 16%
- ☐ Heat exchangers - 7%
- ▨ Instrument & electrical - 7%
- ■ Pressure relief valves - 3%
- ☐ Process piping & valves - 31%
- ☐ Pumps - 26%

4.1 Columns and other process pressure vessels and piping

Corrosion

It is not intended to describe in detail all the corrosion mechanisms that exist within CDUs and VDUs. The following is a list of the most frequently found corrosion mechanisms on these units together with the appropriate API RP 571 reference. Some incidents that arose as a result of undetected corrosion are included.

- **Ammonium chloride corrosion** (API RP 571 – 5.1.1.3): Corrosion that forms under ammonium chloride or amine deposits that form on the inside surfaces at the top of a CDU main column, and overheads piping and condensers.

Operators can reduce the likelihood of those deposits occurring through:

- Ensuring good desalter operation with acceptable (low ammonia) desalter wash water quality.
- Maintaining wash water to overheads systems (if used) at the required flowrates.
- Ensuring that filming amine addition is maintained at the required levels.
- Be aware of the salt sublimation temperature for the unit. This will depend on the partial pressure of the salt species.

- **Hydrochloric acid (HCl) corrosion** (API RP 571–5.1.1.4): This forms in the overheads systems of CDU main columns as the first drops of water condense out, and can also be a problem in VDU ejector/condenser sets.

Operators can reduce the likelihood of HCl attack by:

- Maximizing crude oil tank water separation and draining in the tank farm. Many sites have a policy of allowing a minimum settling time of two days before allowing the tank to be pumped to a CDU.
- Ensuring good desalter operation.
- Maintaining wash water to overheads systems (if used) at the required flowrates.
- Maintaining combinations of neutralizing and filming amines at the required levels.
- Monitoring the pH of the overheads water that collects in the reflux drum/ accumulator boot on a regular, typically once per shift basis, and report any excursions outside a predetermined safe range to plant supervisors.

- **Sulphidation corrosion** (API RP 571 4.4.2.1): This attacks carbon steel and other alloys as a result of the reactive corrosion of sulphur compounds (e.g. elemental sulphur, H_2S and mercaptans) in crude oils. It occurs in the hot sections of the unit [generally above 230°C (446°F)]. The higher the reactive sulphur levels in the crude oil feed, the higher the corrosion. The solution is to ensure correct metallurgy selection for the process conditions.

Process operators should ensure full compliance of operating envelopes to avoid sulphidation.

A severely corroded vacuum residue pump suction line suffered major failure after one of the pumps had been isolated following a minor seal fire. Updraught from overhead fin-fans exacerbated the fire. (11)

A CDU splitter pumparound line failed due to sulphidation corrosion, which had gone undetected despite a piping thickness testing programme. The resulting jet fire impinged on other equipment including overhead fin-fans and was extinguished in 2 1/2 hours. (72)

High temperature sulphide corrosion inside an external CDU overflash line resulted in leakage and fire. Low velocity within this line allowed H_2S at 350 °C to separate out accelerating corrosion. External overflash piping was subsequently removed. (122)

- It was the practice in the past to use *'alonized' low alloy steels* in high temperature, high sulphur, service, where aluminium is diffused into the internal surface of the steel. Much of this type of steel has been removed over time, but on occasions it may still exist. There have been a number of incidents within the industry resulting from the failure of this material in recent years. It should be removed at the first opportunity and careful pipe wall thickness monitoring of this material is advised until removal.

A hot naphtha line from a CDU to a hydrotreater was constructed in 1965 from 'alonized' steel for the high temperature, high sulphur environment. Failure to recognize non-uniform corrosion despite an upstream section being replaced some time before lead the 6″ line to fail catastrophically 30 years later, near the base of the vacuum tower. The resultant torch fire and subsequent fires from leaking flanges and pipe failures burned for approximately ten hours, with extensive damage (see picture showing vacuum tower). (65)

- The effects of failure due to corrosion are clear, but the fate of the products of corrosion, such as iron sulphide scale, can also cause problems downstream with blockage and maintaining flow regimes within their safe operating limits.

Opening of the bypass around the blocked and infrequently used heavy slop oil recycle flow control valve caused overheating in three heater passes followed by a rapid loss of VDU column vacuum and a transfer line flange fire. (27)

- **Naphthenic acid corrosion**—NAC (API RP 571 5.1.1.7): This is a form of high temperature corrosion that occurs in CDUs and VDUs when crude oils that contain naphthenic acids are processed. It mostly occurs in the hot sections of CDU/VDUs at above 200°C (392°F) and in areas of high turbulence. Piping systems are particularly vulnerable in areas of high fluid velocity.

Crude oils are normally considered as naphthenic and requiring special material selection when the TAN (Total Acid Number) is >0.5 mg KOH/gm

The main mitigation method is through the selection of the process equipment metallurgy.

Process engineers and production planners can reduce the impact acid corrosion has on a unit that is not designed to handle it by blending the crude oils to reduce the TAN of the feedstock.

High temperature NAC inhibitors are available, and can be used in conjunction with extensive corrosion monitoring systems to mitigate acid corrosion. The dosage rate must be carefully monitored and controlled in line with the measured corrosion rates.

- **Wet H$_2$S damage** (API RP 571 5.1.2.3): This describes a range of damage that can occur to carbon and low alloy steels through blistering or cracking. The basic chemistry is based on the reaction of H$_2$S with the iron oxides in pipescale that create iron sulphide and hydrogen atoms. The hydrogen atoms diffuse into the steel of the pipe or equipment wall, collecting at a discontinuity or inclusion and then combine to form hydrogen atoms which become trapped because of their larger size causing blisters or cracking.

Mitigation is through selection of materials of construction, including not using 'dirty steels'. Internal coatings are effective, such as Monel, used at the top of CDU main columns. Post weld heat treatment (PWHT) is effective in reducing the effects of wet H$_2$S damage.

Operators can play an important role in testing aqueous streams for pH and reporting any deviations from predetermined safe limits to plant supervisors.

WARNING: The sampling of high H$_2$S streams is hazardous and must be carried out strictly in accordance with safe working practices and procedures. Closed-loop sample systems or the use of Self Contained Breathing Apparatus (SCBA) is strongly recommended.

- **Caustic stress corrosion cracking** (API RP 571 4.5.3): This occurs in carbon and low alloy and 300 series stainless steels exposed to caustic soda (NaOH) or caustic potash (KOH).

Cracking can be prevented in carbon steel piping and vessels by subjecting piping and process vessels to post weld heat treatment (PWHT), which requires heating to 621°C (1,150°F) to relieve stresses created in the fabrication processes.

Particular care needs to be taken with non-PWHT carbon steels with steam tracing design and when steaming out.

Caustic stress corrosion cracking leading to a 76 cm (2.5 ft) long rip in the 20″ main crude line occurred at a caustic injection point relocated from downstream to upstream of the pre-heat exchangers. The injection quill had been incorrectly positioned. (46)

Inadequate design and construction

- **Corrosion within 'dead legs'.** Piping 'dead legs' are not expected to be found in original process unit designs. Areas where no flows are expected to occur, such as at the bypasses around control valves or heat exchangers are designed to be free draining to the main process flow. However, when changes are made to the unit design, dead legs can be formed as sections of redundant piping are positively isolated rather than being removed. See also section 3.1.

Piping failure due to corrosion within a 'dead leg' section of cross over piping on the suctions of the VDU flash zone reflux pumps resulted in major loss of vacuum. A fire and explosion occurred inside and outside of the column. (137)

A fire occurred when the desalter RV discharge piping adjacent to the CDU main column failed due to corrosion. This piping is normally a 'dead leg' and was operating at a higher temperature than that specified for the piping material of construction. (54)

Failure through corrosion of a crossover line, normally a 'dead leg', on the suction side of the vacuum column residue and flash zone pumps released hot hydrocarbon that auto-ignited. The fire was exacerbated by failure of the pump discharge pipework. (19)

Some 'dead legs', however, are formed in piping systems that are only used occasionally. These situations can be overcome by the use of exotic metallurgy or by isolating at the normally operating hot piping and flushing the infrequently used piping with gas oil or similar.

A stagnant line within the CDU/VDU residue systems suffered major failure due to internal sulphide corrosion. This line was only used at start-up and shutdown, i.e. for about two weeks every two years, at which time it operated at 332°C (630°F). (115)

'Dead legs' can also accumulate sludge and chemicals, which promotes aggressive under-deposit corrosion, in this case possibly hydrochloric acid corrosion.

> A stagnant section of a heat exchanger bypass piping failed during a pressure surge within the unit. Significant crude sludge and brine had accumulated where the piping sloped away from the main process piping. (55)

- **Inadequate material of construction specification**. Many major incidents arise from the wrong material of construction being inadvertently used. Failures through corrosion can occur at seemingly insignificant locations and take many years (but can be reduced to a matter of weeks when inadequate materials are used). The consequences of failure can lead to major losses. See also section 3.1.

> A small section of 1/2% Mo carbon steel pipe had been installed between the 5%Cr 1/2%Mo CDU transfer line and a thermowell. The piping slowly corroded over 20 years before failure through corrosion occurred. (12)
>
> A leak occurred on a VDU when a 4″ line that was used to recirculate hot distillate at 370°C (700°F) into the unit feed before the furnace [approximately 250°C (480°F)] ruptured (see picture adjacent). After removal of the insulation, the release auto-ignited, resulting in a serious fire. This 4″ pipe was made of ordinary carbon steel and installed ten years before (140).

Failure can also occur at welds of dissimilar materials, particularly if the wrong materials are used.

> A flange of the wrong material that had been installed on the VDU residue pump discharge line corroded to failure at the weld zone. Further investigation showed many other examples of incorrect material usage in the residue piping system. (96)
>
> An error in defining the correct piping material specification break had led to carbon steel pipe being used where 5Cr/0.5Mo was the correct specification for a VDU residue recycle line to the charge heater inlet, leading to corrosion and failure. (128)

Elbow from a VDU with wrong metallurgy that failed after eight years resulting in an aerosol leak at column bottom and a small fire. (144)

It is recommended that every plant have a material verification programme for both existing equipment, new equipment, and for replacement and reassembled components (refer to API RP 578 *Material Verification Program for New and Existing Alloy Piping Systems*).

- **Inadequate supporting**: Inadequate supporting has been the source of many incidents, either by vibrations, pipes falling off supports or by excessive stress on welds.

An operator noticed a naphtha smell during a routine inspection. After the removal of insulation, a 40 cm (16-inch) long crack was discovered along the gusset weld where a pipe support was joined with the crude pre-flash column (See picture below). There was a potential for a major fire so the unit was shut down for inspection and repair. This incident shows the importance of regular operator walkabouts to detect unusual 'happenings'.

- Inadequate design—pipe support was welded directly to the column without any reinforcement plate.
- Poor welding—lack of weld penetration.
- Fatigue failure—Support bracket vibrated due to two phase flow regime in the pipe. (142)

Crack

4.2 Desalters

Critical operating parameters impacting process safety

Critical operating parameters are included in refining design manuals and individual process unit operating instructions. Some of the critical parameters that plant supervisors and operators need to be aware of are listed below:

Operating personnel should regularly monitor salt, solids and the water content of the incoming crude mix and desalted crude oil. Other parameters should also be monitored frequently:

- crude oil salt content;
- solids and BS&W content;
- water quality and injection rates;
- mix valve pressure drop;
- desalting chemical injection type and flowrates;
- grid voltage and amperage;
- desalter temperature and pressure;
- brine pH and the level of the brine/crude oil interface;
- the quantity of slop oil where this is injected into crude.

WARNING: Sampling of hot brine is hazardous and must always be carried out using a sample cooler, and wearing a full face shield and impervious gloves.

The quality of aqueous effluent from desalters

This must be regularly monitored. Sight glasses that allow operators to monitor the colour of the effluent are particularly vulnerable to failure if not correctly specified and fitted. It is essential that the anti-corrosion films made of transparent mica are correctly installed on the inside surfaces of these sight glasses.

A 'bulls eye' sight glass failed on the brine outlet from the desalter. The replacement unit also failed. Both failures were attributed to insufficient thickness of glass for the pressure duty. (56)

Desalter water supply

Part of the water supply arrangements to desalters are normally fed from an atmospheric cone roofed tank located on the unit. There have been a number of incidents where the water injection pumps have failed and back flow prevention, non-return valves and/or low flow trips, has not operated. The consequences are that crude oil may back flow into the desalter water tank and release light ends from the tank vent. There is also a risk of overflow from the tank.

The coupling failed on a CDU water injection pump resulting in back flow of crude oil into the water injection tank as two non-return valves (NRVs) or check valves on the pump discharge failed. The tank overflowed with crude ignited by sparks from the broken pump coupling. (4)

A NRV failed to prevent back flow when a wash water pump tripped. The wash water tank overflowed crude oil into the sewer system. In this case no ignition occurred. (14)

Electrical supply to desalter electrodes

Great care must be taken when maintaining the high voltage electrical supply transformers and their associated transformers, cabling and electrodes.

A contractor electrician received a high voltage electrical shock when working on electrode No: 1 at the desalter. He was scheduled to work on the isolated No: 3 electrode. An ambiguous drawing was displayed in the control room and no tag number check was made. (81) See also section 6.2.

Vibration within relief valve pipework

Damaging vibration can be set up in relief valve piping where the pressure drop across the inlet piping between the main process inventory and the RV inlet nozzle is greater than 3% of set pressure. The phenomenon is aptly called 'machine gunning'. Desalters may be prone to this as it is usual to locate the desalter RVs at the CDU main column. This allows the RVs to discharge directly into the column flash zone, and avoid the need for large bore piping runs that would be required to carry two-phase flow to the main column if the RVs were to be fitted directly onto the desalter vessel.

The requirement for RV inlet pipework pressure drop to be less than 3% of set pressure is contained in ANSI/API 521 *Pressure-relieving and Depressuring Systems*.

The internal bellows of a balanced bellow RV failed through vibration when the desalted crude oil pump tripped raising the desalter pressure above its set pressure. No ignition occurred. (49)

Similar vibration to that described above was observed when desalter RVs lifted discharging directly into the main column flash zone. Calculations carried out after the event showed that the pressure drop in the piping between the desalter and RVs was well in excess of the 3% of set pressure (as required by API RP 521 to prevent 'machine gunning') when the RVs were fully open. (138)

A fire occurred in a CDU during start-up after a four week maintenance turnaround. Due to excessive leakage from two heat exchangers it was decided to stop the start-up to rectify these leaks. When the feed rate was reduced, the pressure in the desalter increased inadvertently and the desalter RV opened and relieved into the crude column. There were severe vibrations in the RV and related piping, and a flange leak started at the RV by-pass valve (see arrow in picture below). This ignited within seconds, and large flames erupted next to the crude column. The fire was extinguished after about two hours. The unit was shut down for a further four weeks to complete necessary repairs. (141)

4.3   Fired heaters

BP Process Safety Booklet *Safe Furnace and Boiler Firing* gives advice on critical safety and operational issues.

Critical operating parameters impacting process safety

- Fired heaters provide the primary source of heat for the atmospheric and vacuum distillation of crude oil. Oil, preheated in heat exchangers, is heated in the charge heater, or pipestill, to the temperature required to vaporize the required distillates. The maximum temperature to which the oil can be heated is an important variable in fired heater operation since excessive temperature increases the rate of coke formation, and its associated cracked gas generation, within the heater tubes. Good fired heater operation and a knowledge of the factors that indicate or influence tube coking are of prime importance to ensure safe and efficient operation between turnarounds. When assessing the safe operation of fire heaters operators need to take account of *all* these parameters together.

Supervisors and operators must be fully aware of the maximum tube skin temperature allowed in a fired heater. This is normally monitored by skin thermocouples, but these can be prone to failure and can only be repaired at turnarounds. When they fail they usually go to full scale. However, full scale skin temperature readout is also indicative of a much reduced flow within the heater tubes caused by some external factor such as a fault with the heater tube pass flow measurement device and/or control valve. Any tube skin temperature excursion towards full scale, therefore, must be immediately investigated to ascertain the true cause and the fired heater shutdown activated without delay if a reduced or no flow condition is found. Tube skin temperatures can occasionally be checked using infra-red thermography.

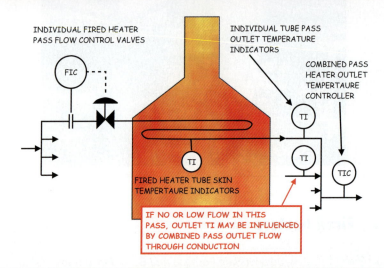

INDIVIDUAL FIRED HEATER PASS FLOW CONTROL VALVES

INDIVIDUAL TUBE PASS OUTLET TEMPERATURE INDICATORS

COMBINED PASS HEATER OUTLET TEMPERATURE CONTROLLER

FIC

TI

TI

TI

TIC

FIRED HEATER TUBE SKIN TEMPERTAURE INDICATORS

IF NO OR LOW FLOW IN THIS PASS, OUTLET TI MAY BE INFLUENCED BY COMBINED PASS OUTLET FLOW THROUGH CONDUCTION

Although not a CDU or VDU fired heater, a fired reboiler on a fractionation tower showed signs of excessive skin temperature on one of the heater passes. This was assumed to be an erroneous reading. What happened was that a fault in the pass flow controller had resulted in the pass flow control valve gradually moving to the closed position while the flow measuring device outputted a constant value. The condition was detected by visual inspection of the tube pass, which had turned a cherry red colour. The unit was shutdown immediately, but when opened up, the cross section area available for flow in the 8" tube was effectively reduced by three quarters.

- Steam injection or the use of a recycle hydrocarbon stream must be used to maintain heater process tube velocities when the feedrate typically reduces to 60% of design or below. If this is not done, enhanced coke formation on the internal surfaces of the tubes may result in thermally insulating the metal of the tube wall from the process fluids. This may cause overheating of the tube wall leading to enhanced corrosion and scale laydown on the outside surface and bulging of the tube wall, ultimately leading to failure.

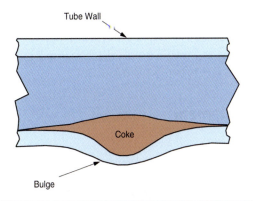

Tube Wall

Coke

Bulge

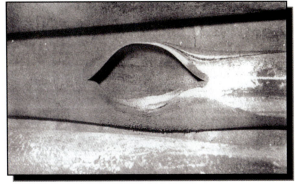

- It is essential that the operating range of the heater has been checked and confirmed by a fired heater specialist, and is well understood by supervisors and operators.

Balanced firing is essential to prevent premature tube overheating and coking/fouling. A 55°C (100°F) difference between the highest and lowest firebox temperature readings is acceptable (not including end walls), with 110°C (200°F) being the maximum allowed.

Heater pass outlet temperatures should be balanced within 3 to 5°C (5 to 9°F). However, the variation in flowrates between the various passes should also be monitored and generally should not vary by more than ±10% of the average.

Isolation of fuel supplies

As discussed above, any major reduction in the flow through the tube within a fired heater usually requires immediate action to shut off the main fuel flows to the heater—the pilots can be left in commission. Because of the higher temperatures in VDU heaters, emergency steam injection into the tube pass inlet must also be commissioned.

Naphthenic acid corrosion

Naphthenic acid corrosion (API RP 571 5.1.1.7) is described in section 4.1.

Heater tubes and the transfer lines between the heater and the main CDU or VDU column are particularly vulnerable to NAC.

4.4 Rotating equipment

Corrosion

- Similar problems to those experienced in columns, vessels and piping apply to pump casings where a screwed plug may be replaced without any thought being given to the need to ensure it is of the correct specifications.

> A carbon steel plug was ejected from the body of a hot gas oil pump at 380°C (716°F) at a VDU causing a major fire. The plug had been fitted in error and had suffered from corrosion; the correct specification was 13% chromium steel alloy. (80)

Vibration

Small bore connections to pieces of unsupported equipment, such as pressure gauges are extremely vulnerable to fatigue failure. The problem becomes exacerbated:

- on equipment that has a natural vibration cycle, such as reciprocating compressors;
- on equipment operating at its critical speed;
- at the higher pressure ratings as isolation valves as well as the equipment itself becomes increasing large in respect to the mechanical strength of the connecting piping.

> It is likely that vibration led to the failure of a manometer nozzle on a CDU overheads gas compressor. (26)

- **Some small bore piping connections**, such as some compression fittings are unsuitable for severe hydrocarbon or critical utility duty. This type of connection will usually not be found in the unit piping specification, but may be provided as part of a vendor package, as in the example below. In some cases vendor supplied packages contain general purpose items that do not meet the strict requirements of the codes and standards used in the petroleum industry, such as API.

> Leakage from a light compression fitting on a seal oil return line was ignited by the casing of a residue pump. This type of compression fitting was vulnerable to vibration and not recommended in most engineering standards. (15)

- **Sometimes corrosion and vibration act together** to cause major loss of containment, particularly when a component does not meet the material specification for the duty.

> A screwed nipple, part of a heavy gas oil reflux pump vent, failed through corrosion and vibration. The nipple was below specification thickness and failed where the screw thread was cut. Hot oil escaped and auto-ignited. (132)

Mechanical seals and seal systems

> The API Mechanical Equipment Standards for Refinery Service give advice on all aspects of rotating equipment used in refineries, including seal system options. These include: API 610 *Pumps for Refinery Service*, API Standard 617 *Axial and Centrifugal Compressors and Expander-compressors for Petroleum, Chemical and Gas Industry Services* and ANSI/API 618 *Reciprocating Compressors for Petroleum, Chemical, and Gas Industry Services*.

- **Mechanical seals** replaced packed glands on petroleum refining pumps in the late 1960s. Early designs did not have any secondary sealing with some requiring cold seal oil to keep the seal faces below the temperature at which they failed (this safety critical flow was often supplied by an offsite pump with little or no redundancy provided). Seal faces were prone to coke formation, with mitigation sought through placing a steam lance on the leaking seal. In the case described below, which is typical of that era, it was clearly an ineffective solution. On many CDUs and VDUs built on existing plots where space was limited, the congestion of equipment around and above hot pumps added to the damage caused when a seal failed.

> An atmospheric residue pump mechanical seal failed engulfing the pump filter in fire and destroying instrument and electrical cables. A short shutdown resulted to replace damaged cabling. (6)

- **Bearing failures**. The satisfactory operation of all rotating equipment seals is dependent on the rotating elements running true. Any form of bearing failure can destroy the seal leading to major loss of containment. Where hot pumps are involved the escaping process fluids may auto-ignite. However, there is sufficient frictional heat generated by rotating and static components coming into contact with each other through misalignment when a bearing failure occurs to ignite the release.

> A large fire occurred as the result of seal failure on a 'cold' crude oil pump [130°C (266°F)] that was caused by a pump bearing failure. The escaping crude oil was ignited by friction generated at the failed bearing. (79)
>
> Seal failure of a top pumparound pump was most likely ignited by frictional heat within the seal. (20)

- **Seal oil systems.** Mechanical seals require a supply of cool, clean fluid at the seal faces. Many pumps have a simple supply system routed from the discharge of the pump to the seal assembly that passes through a cooler and small cyclone that removes any solid particles. This type of system is normally isolated with the pump itself and requires no intervention from the operator except to ensure that cooling water is available to the cooler at all times.

  A more complex system for some mechanical seals is to supply seal oil from an external source. Where this occurs the supply system must be considered a safety critical system and designed as such. If the seal oil supply fails in any way the pump must be shutdown immediately.

  Seal oil systems for tandem and double mechanical seals are also complex and they must be considered to be safety critical systems. See point on mechanical seal design below.

- **Isolation of hot oil pumps in an emergency.** Operators isolating hot pumps after a seal had been found to be leaking reported a sudden escalation of the leakage as the problem pump was being isolated after the spare pump was started. This could arise where the warm-up bypass of the problem pump had not been closed after it had been previously started. Warm-up bypasses connect the discharge line from the pumps back into the pump body, which can achieve discharge pressure when the suction valve is shut. Seals normally only see the pump suction pressure. Escalation of leakage can also result as the pump speed runs down.

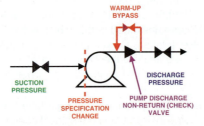

A major fire resulted from an escalation of leakage from an atmospheric residue pump mechanical seal. The pump was being isolated following observation of a slight, unignited, release. The spare pump had just been started up. (12)

Similarly an increase in pressure can cause flange leakage on the suction side of the pump.

An ignited release from an atmospheric residue pump seal was quickly extinguished using dry powder, but a flange leak occurred on the pump suction filter as the pump was being isolated and reignited the original release. (10)

- **Consequences of seal failure** of any pump or compressor that suffers seal failure from whatever cause can have a major impact on adjacent or overhead equipment. Unprotected instrument and electrical cables are particularly vulnerable. Cables and adjacent equipment can be protected in a number of ways.

  o Relocating hot pumps away from areas below overhead equipment, which may not be an option on an existing plot.

  o Installing instrument and electrical cables that are inherently fire protected, which is expensive.

  o Installing passive fire protection around cables which are not inherently fire protected. Where this is done it is essential that the integrity of the cable protection systems is maintained at all times, with covers removed for maintenance and inspection replaced as soon as inspection work is completed or suspended (e.g. overnight).

> Bearing failure on a kerosene pump resulted in seal leakage that was ignited by the hot bearing. No vibration or high temperature alarms were fitted to the pump. Neither was there any fireproofing of instrument/electrical cables in the hot pump area. (105)

- **Continuous monitoring of rotating equipment**. In the example above there was no continuous monitoring of the rotating equipment. The cases described below add strength to this argument. Where such systems are installed and for some reason fail to operate, this must be taken to be a failure of a safety critical system and acted upon accordingly. These alarms must never be ignored.

> Failure of the inboard bearing of a steam turbine driven CDU atmospheric residue pump caused a major seal failure before operators could change over to the spare pump. Escaping residue was ignited by auto-ignition. (37)

> Mechanical seal failure at the remote cold crude charge pumps resulted in the CDU shutdown. An alarm set off by vibration at the pumps was ignored by CDU control room operators as it was considered unreliable. (63)

- **Mechanical seal design** applied to safety critical duties generally contains double containment features, such as double mechanical seals and secondary seals.

  It is generally recommended that double mechanical seals, or equivalent, be installed on:

  o Hot pumps operating at temperatures greater than the auto-ignition temperature;

  o LPG pumps;

  o Pumps with the suction under vacuum;

  o High H_2S duty, i.e. could result in a release of more than 1,000 ppm into a breathable atmosphere.

49

These designs can include monitoring of seal conditions and the supply of sealing fluids, steam cooling etc. The description below demonstrates the consequences of a component failure within a single mechanical seal with no enhanced containment features.

Fluid release through a cracked bellows inside the mechanical seal of a crude oil booster pump ignited and escalated into overhead equipment, resulting in a CDU shutdown. The investigation recommended remote isolation and secondary sealing for critical pumps. (118)

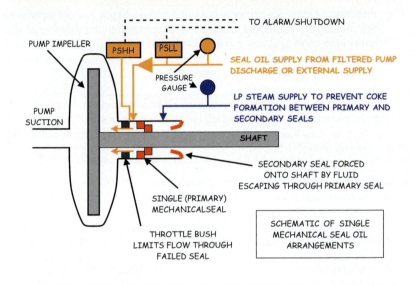

TO ALARM/SHUTDOWN

PUMP IMPELLER

PSHH

PSLL

PRESSURE GAUGE

SEAL OIL SUPPLY FROM FILTERED PUMP DISCHARGE OR EXTERNAL SUPPLY

LP STEAM SUPPLY TO PREVENT COKE FORMATION BETWEEN PRIMARY AND SECONDARY SEALS

PUMP SUCTION

SHAFT

SECONDARY SEAL FORCED ONTO SHAFT BY FLUID ESCAPING THROUGH PRIMARY SEAL

SINGLE (PRIMARY) MECHANICALSEAL

THROTTLE BUSH LIMITS FLOW THROUGH FAILED SEAL

SCHEMATIC OF SINGLE MECHANICAL SEAL OIL ARRANGEMENTS

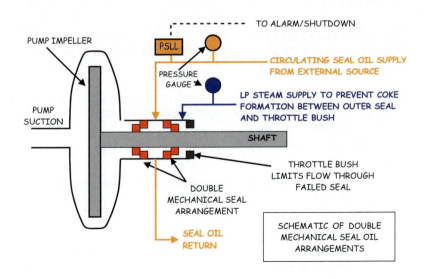

TO ALARM/SHUTDOWN

PUMP IMPELLER

PSLL

PRESSURE GAUGE

CIRCULATING SEAL OIL SUPPLY FROM EXTERNAL SOURCE

LP STEAM SUPPLY TO PREVENT COKE FORMATION BETWEEN OUTER SEAL AND THROTTLE BUSH

PUMP SUCTION

SHAFT

THROTTLE BUSH LIMITS FLOW THROUGH FAILED SEAL

DOUBLE MECHANICAL SEAL ARRANGEMENT

SEAL OIL RETURN

SCHEMATIC OF DOUBLE MECHANICAL SEAL OIL ARRANGEMENTS

Remote isolation facilities for rotating equipment

In the above example reference is also made to remote isolation of pumps containing process fluids that are flammable, will auto-ignite or are toxic. These generally take the form of an actuator fitted to the suction valve or a separate emergency shutdown valve fitted to the suction line as it leaves the inventory of hazardous materials. The actuator itself and any instrument and electrical cable is fire protected, with the actuation point at least 30 metres away in a safe location, and/or in a remote control room. The UK Health & Safety Executive has published contract research report 205/1999 *Selection Criteria for the Remote Isolation of Hazardous Inventories* which gives guidance on this type of installation.

Couplings

- Couplings are complex multi-element devices that need careful attention in selection and installation.
- Misalignment can occur if the measurements made by the installing technician are not accurate or the driver and driven equipment move out of alignment over time, something that should not be possible if the respective pieces of machinery are properly located to the baseplate with dowels. When a coupling does fail it can cause failure of other components leading to loss of containment.

The sudden failure of the coupling on a CDU intermediate reflux pump caused the mechanical seal to fail releasing 290°C (554°F) hydrocarbon that ignited. The fire was drawn up through the unit piperack by the updraught from fin-fans located above. (97)

- Another example of the consequences of misalignment.

A CDU atmospheric residue pump was severely damaged when the cast iron support feet failed, possibly as a result of the motor being forced into alignment creating movement within the clearance between bolt and baseplate threads (102)

- Couplings are also vulnerable to changes made to any part of the rotating equipment that influences critical speed.

A major change occurred to the critical speed of an atmospheric residue pump that had been converted from packed glands to mechanical seals. The pump eventually failed catastrophically. (100)

4.5 Heat exchangers

Critical operating parameters impacting process safety

- CDU and VDU heat exchangers, including condensers, coolers and reboilers, play an important role in allowing the units to operate in an energy efficient manner. Exchangers perform two functions—they maximize the transfer of heat from products to feedstock minimizing the amount of heat required from external sources, and they control the amount of heat lost to air or water in cooling products.

- Factors that control good exchanger operation and performance include: feed type, salt, sediment, contaminants, desalting performance, tube-side velocity, exchanger design, corrosion, temperature, water-side pre-treatment, etc. Even at low temperatures, corrosion and fouling can have an important impact although the operator has no control over the treatment of the raw crude.

> Operators and engineers should maintain records of pressure drops across all exchangers and monitor closely those exchangers that are known to be in fouling duty.

Heat exchanger fouling

- Heat exchanger performance and fouling should be monitored regularly. Failure to transfer the required amount of heat can cause fired heaters or other equipment to become overloaded, further hampering efficient and safe operation and, in extreme cases, resulting in equipment failure.

> Heat exchanger tube side velocities should generally be maintained between 1.8–3.0 m/s (6–10 ft/s) to reduce the likelihood of fouling.

> Operators should be aware that bypassing and throttling back the flow through a heat exchanger can increase fouling. Also, for cooling water service, cooling water needs to be maintained below 50°C (120°F) to prevent excessive water side fouling—see below.

- **Various methods are used to limit fouling** within exchangers including periodic online cleaning (with Light Cycle Oil and other solvents); offline cleaning (water jetting); and online use of chemical additives. Cooling water exchangers are also periodically back flushed.

- **Heat exchanger fouling** is generally an ongoing problem on CDU/VDUs, particularly on:

  o Overheads systems where a complex mixture of ammonium chloride salts can be deposited.

o   Residue/feed heat exchangers due to asphaltene and coke laydown. Positive isolation and hard piped flushing oil and drain down systems are necessary to ensure that individual or pairs of heat exchangers can be safely bypassed and isolated to allow high pressure water jetting to be carried out.

> Advice on mitigating the hazards of high pressure water jetting is given in Section 3.2. High pressure water jetting should not be used to clean the outside surfaces of fin-fans as this can cause damage to the fins.

> Operators should be alert to the possibility of pyrophoric scale igniting coke deposits when heat exchangers are opened up to atmosphere. *Debris released by a cleaning process should not be allowed to dry out.*

- **Back flushing of shell and tube coolers and condensers** should be carried out where warmer seasonal water may give rise to fouling.

> Outlet water temperatures should not be allowed to rise above 50°C (122°F) to prevent tube side fouling from calcium carbonate deposition. Discharge of water from back flushing to soft ground should be avoided—it should be routed to the oily water sewer

- **Care must be taken when cleaning heat exchangers constructed of exotic materials**

> Scratching titanium tubes while cleaning by rodding or high pressure water jetting can lead to tube failure from hydrogen embrittlement when the equipment is returned to service.

4.6   Distillation column overfilling

Engineering approach and standards

Loss of level control may occasionally occur on CDU/VDUs. This could happen due to many reasons including feed variations, blocked residue stripping trays, blocked level measurement device nozzles, faulty or misleading level indication. Overfilled columns and vessels are a serious safety hazard, particularly if there is a possibility that the column or vessel pressure relief valves could relieve hydrocarbons to atmosphere. Engineering practices should aim at delivering a clear level indication, with parallel redundant level transmitters, in all distillation columns.

In addition to control and relieving issues, overfilling of some columns and vessels may overstress supporting foundations and vessel skirts where these have not been designed for the liquid full condition.

A fire occurred on a distillation crude column as the result of the release of the stabilization column overhead relief valve. The RV lifted because of excessive internal pressure and filling of the column with liquid. The relief valve outlet was routed to atmosphere and was located at the top of the distillation crude column, which was the highest point of the unit. The released hydrocarbons ignited on a hot point at the foot of the column (see picture).

The damages were moderate and affected mainly instrumentation cables and insulation. (143)

Guidelines for all column and vessels subject to liquid overfill

- All columns and vessels (new-build or old) with Atmospheric Relief Valves (ARVs) require two separate and diverse methods of level measurement. An exception is where existing vessels/drums are normally operated with zero liquid product (e.g. compressor knock-out pots).

- For existing equipment with Pressure Relief Valves (PRVs) to a closed flare, retrofitting to two separate and diverse level transmitters is highly recommended.

- Critical alarms are required on all columns and vessels. Critical alarms may be taken from the non-controlling transmitter (no requirement for an additional separate hard-wired system).

- Hazard assessment [or first part of Layers Of Protection Analysis (LOPA)] for liquid overfill is required for all columns and vessels. This will determine whether further instrumentation or a high-level trip is required.

- The order of preference for level transmitters is guided wave radar, nuclear, DP, magnetic coupled.

- New-builds need separate level transmitter nozzles, but retrofits may use existing single tappings depending on plugging history.

5

Safe operating practices and procedures

5.1 Safe operating practices

- Safe operating practices cover the things that operators do where they are not expected to refer to a written procedure every time they carry them out. In the majority of cases, safe operating practices are described in detail in the training manuals for each process unit. *These manuals should always be kept up to date* as the unit and procedures are modified. An example could be starting and stopping a process unit pump that has low complexity in its design or duty. This is in contrast to operating multi-stage pumps that deliver high pressure and require minimum flow to be maintained at all times; these will require a dedicated *procedure* that is closely followed every time the pump is started and stopped.

- Operators are required to know how to respond to an emergency scenario without having to refer to detailed operating instructions. These scenarios include utility failure, loss of feedstock etc. Again, the procedures are detailed in operating and training manuals, but it is critical that responses to emergency scenarios are practised regularly through short 'gun drills' (*so called after the frequent gunnery exercises carried out by the Royal Navy that allowed them to outgun their opponents—a matter of survival*).

It is considered good practice to summarize the critical operating steps on a prompt board located at the equipment itself. For example the actions taken when a flame out occurs on a fired heater. The full procedure will be in the emergency shutdown section of the operating manual, but where delay in accessing the full manual may exacerbate an unsafe situation a prompt board can give confidence to operators that they have not overlooked anything.

Another option is to give pocket cards for each operator position. Each card is printed with the primary shutdown steps for each defined emergency situation (for example, loss of air, nitrogen, steam, etc.).

5.2 Safe operating procedures

Safe operating procedures typically fall into four categories:

Start-up procedures

- These set out a detailed sequence of events that are required to take a process unit from mechanical completion to operating at its design conditions. Each step is described, and space provided for operators to record timings and any unforeseen events.

- Start-up instructions should be written in sections so that each section starts and ends at a safe, stable operating condition. For example, a section could cover establishing levels and cold circulation, or moving from hot circulation to placing the unit on stream.

- While most start-up procedures may start from the point of mechanical completion handover, many will also start from the situation that results from an emergency shutdown or where the unit has been shutdown but not de-inventoried, possibly due to lack of feedstock. *The correct procedures must always be used.*

Shutdown procedures

- As with start-up instruction, these cover the sequence of events required to safely move the operating unit to a shutdown situation.

- The end point may be completely emptying the unit of all hydrocarbon, including purging/steaming etc., or to maintain part or whole of the inventory while some specialized activity takes place, such as fired heater decoking.

- Each section of the procedures should start and end at a safe, stable operating condition.

Normal operating procedures

- These primarily cover how the unit responds to change within the safe operating envelope—see below.

- A critical area is the rate of change that can be made to operating conditions. CDUs and VDUs are highly thermally integrated units and even small changes made to a temperature or flowrate can have a wide-ranging impact on the unit, which takes time to fully work through the system. Rapid or frequent changes can create instability, which can manifest itself in a number of ways, including:

  o Sections of the main columns flood or run dry leading to tray damage or pump damage.

  o Temperatures may be forced above the safe operating envelope, for example, if firing is increased too rapidly, tube skin temperature may rise to the point where excessive coking takes place and damage occurs to the tube metallurgy, firebox and convection section refractory temperature limits are exceeded, and heater draughts go positive.

- Normal operating procedures should also describe how the unit responds to change made within the safe operating envelope. For example, increasing or decreasing a side stream draw off will impact the boiling ranges of the products both above and below the affected offtake. Raising the column pressure will increase offgas flowrates, and raising the column top temperature will decrease offgas flowrates.

- Normal operating procedures should also cover situations that may not occur for most of the time, such as winterization to protect against low ambient temperatures, and excessive summer ambient temperatures. It is insufficient to write generic procedures—these procedures need to be detailed and specific to pieces of equipment particularly piping systems.

Emergency procedures

- Emergency procedures are generally written against a number of foreseen scenarios, such as electrical power failure, other utility failure, feedstock not available.

- As discussed above, operators will not have the luxury of time to consult the written procedures when an emergency occurs (they may have time to go back to these later to check that everything has been done and in the correct order), so it is necessary to have these embedded in the minds of those who have to respond to them. The frequent use of 'guns drills' and the siting of prompt boards are essential aids to ensure that the emergencies are responded to in a manner that does not make the situation worse.

5.3 Troubleshooting

Major plant upsets can be complex to resolve, particularly when the causes are not immediately apparent. The incident described below was such an event, which demonstrates that it is sometimes clearer to those who are not immediately involved in trying to put things right than those who are trying to do just that. A similar situation arose at the Three Mile Island nuclear power accident in March 1979 where it is alleged that senior operators from an adjacent control room obtained in a short time a clearer picture of what had happened than those who had been there from the start.

A major VDU plant upset occurred as a result of internal leakage within a crude oil/vacuum heavy gas oil reflux heat exchanger. VDU main column RVs lifted. It took twelve hours to troubleshoot a complex scenario to a successful conclusion. (85)

5.4 Concept of safe operating envelope

There are different definitions and terminologies used in the concept of safe operating envelopes (SOE), so it is important to understand the ones used here.

The SOE covers the range of process parameters that will allow safe operation of the process equipment and associated piping systems. The simplest example is that of a liquid level that is controlled within a range determined by the designers, say 20–80% of the control range. When the liquid level lies within this range, operations can be conducted safely. If it moves outside of this range, then unsafe conditions may result. It is for this reason that alarms may be set to warn operators that the level is moving towards an unsafe area, set at say 30% and 70% of the control range. If the operator does not respond sufficiently to prevent the level from reaching the limits of the SOE, then an independent level trip may be provided to take executive action, i.e. to initiate a partial or full plant shutdown. A pressure relief system may be the final device to preserve mechanical integrity of the equipment and piping.

More complex SOEs may combine pressure and temperature, for example, where because of embrittlement problems it may not be possible to increase the pressure of a vessel until a minimum temperature has been achieved.

A simple diagrammatic representation is shown below:

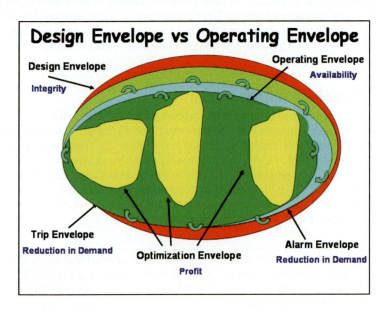

- The DESIGN envelope represents the design pressure and temperature of equipment, relief valve setting, High Integrity Process Shutdown systems (HIPS or 'category 1' trips). The DESIGN envelope contains an engineering design safety margin which can be eroded over time. These are INTEGRITY issues.

- The TRIP envelope is there to reduce demand on the relief system, except where Category 1 or Safety Integrity Level (SIL)—2 instrumented protection systems are the primary protection.

- The ALARM envelope reduces demand on the trip system. The key point is that the operators have time to respond to an alarm to bring the plant back into a safe operating area.

- The OPERATING envelope is there to achieve a given plant availability—it is the deterioration mechanism that it identifies—for example, high furnace temperatures reduce tube life.

- The OPTIMIZATION envelopes drive the plant to operate profitably. They are within the operating envelope and move with operating conditions. As long as the optimization envelopes stay within the operating envelope, changes do not need to go through a Management of Change review for process safety.

- *It is important to note that a Management of Change (MOC) review needs to be performed for all but Optimization Envelope limit changes (as long as these Optimization changes stay within the Operating envelope).*

6

Some serious incidents that have occurred on CDU/VDUs

The following are a summary of some serious accidents that have occurred on CDUs and VDUs.

6.1 Fire and fatalities at Crude Unit tower

ACCIDENT As part of a major overhaul, a CDU was shutdown and made ready for maintenance work. A hot work permit had been issued for the sections of the unit which had been isolated and were considered to be essentially hydrocarbon freed. However, the 1250 mm (50 inch) diameter nozzle on the main fractionator connecting to the overhead line had not been blinded.

A coordination meeting was held on the day before the incident. The intent to carry out four hot work jobs was verbally agreed, but on the following day an application for authorization for additional hot work for repairs on the overhead line of the main fractionating tower was made. Although a hot work permit had been issued for this section of the unit, the work permit system required a separate authorization for each individual job before work could proceed. This authorization was routinely delegated during the turnaround period by the responsible operator to safety officers, who may be safety specialists or suitably trained firemen. A safety officer approved the application for hot work on the overhead line after gas test results indicated that no flammable atmosphere was present.

The maintenance contractor employees then began the hot work on the overhead line of the main fractionating tower, cutting out a coupon from a relief valve branch prior to welding on a new connection. At the same time, smoke and some flames were seen at the top of the tower.

Inside the tower, a scaffolding contractor working on the top tray made a speedy exit because of the incoming smoke, but two operators inspecting the top section of column trays did not escape. Rescue operations were hindered by the large amount of smoke coming out of the upper manway doors. When the smoke from the manways ceased, the two fatalities were recovered from tray 12.

(continued)

The responsibility for authorizing the hot work had been delegated to a safety officer who apparently had no overview of the work in progress or the hazards involved in this particular job. It was not recognized that hot work at a remote location on a line, even though it had not been positively isolated from a vessel, could present a risk to those working within the vessel. Further, although gas tests confirmed that the line was gas-free, there was no guarantee that the lines were clean of deposits or flammable liquids below their flash points, and no extra precautionary measures were taken to ensure that this did not represent a hazard to work.

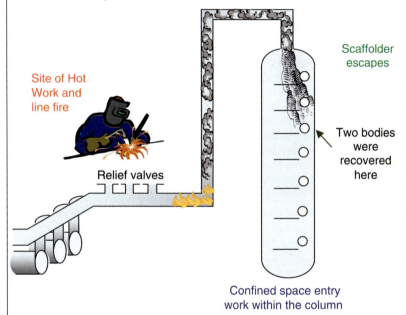

Site of Hot Work and line fire

Relief valves

Scaffolder escapes

Two bodies were recovered here

Confined space entry work within the column

Those who give final authorization for hot work should be sufficiently trained and suitably aware to competently assess all the hazards and risks. This requires full knowledge of the work in progress within the area concerned. It is therefore preferable that the responsible operator should be directly involved in the issue of every authorization, although it is recognized that this may not always be possible given the high number of permits and work activities during a turnaround.

The level of authority for approval to proceed with work must match the level of risk involved.

6.2   Electrocution incident on CDU desalter

Desalters, installed on CDUs and VDUs are unusual pieces of process equipment as they contain internals that are deliberately made live with high voltage, high frequency electricity.

ACCIDENT

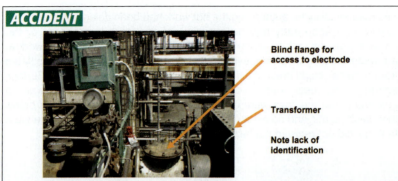

Blind flange for
access to electrode

Transformer

Note lack of
identification

An electrician received an electric shock from a 20 kV supply while he was working on a live electrode on the desalter of a distillation unit. The electrical equipment used in the desalter had only one pull-out circuit breaker to protect the three transformers, each of which supplies one electrode grid. In order to continue operating the desalter with the two remaining electrodes, another electrician had disconnected the electrical supply circuit to the faulty electrode's transformer (#3) on the circuit breaker side in the substation. The electrician actually went to work on electrode #1 because there was a lack of identification on the three transformers; and because the mimic panel available in the control room showed the transformers in the order 1, 2, 3, when in fact they were in the order 3, 2, 1 when looking from the main access route.

Fortunately, he was not fatally electrocuted as his hand was in contact with an earthed/grounded part of the equipment at the moment at which the tool that he was holding touched the live cable. He suffered severe burns on his hand.

The investigation concluded that the main contributing factors were as follows:

- Inadequate checks/tests to determine that the electrical conductor to the electrode was isolated.

- Inadequate identification/tagging/labelling of the electrical equipment.

The fact that there was only one circuit breaker for three transformers was also listed as a design anomaly but it did not contribute directly to the incident.

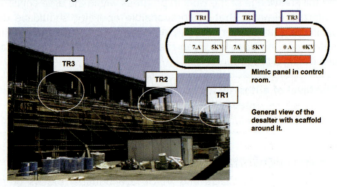

Mimic panel in control
room.

General view of the
desalter with scaffold
around it.

Note: Mimic panels are graphical representations for control purposes. Layout drawing and equipment databanks must be referenced and such information kept up-to-date.

6.3 Hazards of water entering vacuum towers

Water flashing into steam in a vacuum system creates a far greater hazard than is found in the start-up of other types of units. This is because the volume expansion upon vaporization is much greater due to the low pressure in the system. For example, water expands in volume nearly 5,000 times upon vaporization at about 71°C (160°F) at 20 inches (50 cm) mercury vacuum, which is about three times the expansion rate of when water flashes to steam at atmospheric pressure.

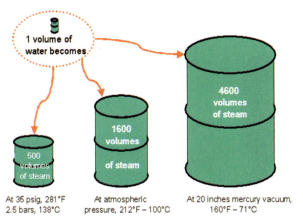

1 volume of water becomes

500 volumes of steam — At 35 psig, 281°F 2.5 bars, 138°C

1600 volumes of steam — At atmospheric pressure, 212°F – 100°C

4600 volumes of steam — At 20 inches mercury vacuum, 160°F – 71°C

A small amount of liquid water can expand to a very large volume of steam when heated.

Furthermore, the large diameter of vacuum towers makes the column internals vulnerable to mechanical damage when subjected to the sudden volume expansion resulting from rapid vaporization.

Once the system is under vacuum, water can no longer be drained from low points. If draining is attempted, air rushes into the tower and may form a flammable mixture with hydrocarbon vapours that in turn may be ignited by pyrophoric deposits.

Pulling a vacuum does not evaporate water left in the system unless the temperature is high enough to boil the water. The water must be removed by circulating oil, which is gradually heated, until all water is vaporized and eliminated. Carefully developed procedures are available to accomplish this task.

ACCIDENT All trays in a large vacuum tower were displaced upwards and then dropped to the bottom of the tower during a start-up. The tower had been properly drained, evacuated and then charged with oil. Cold oil circulation had been started through the tower, and the temperature gradually had been raised to about 260°C (500 °F). At this time, normal flow of oil was to be established by opening a valve at the bottom of the tower to line up suction to a pump.

(continued)

Before this particular start-up, a second valve had been installed in the line to the pump to provide a double block valve arrangement to improve isolation for pump maintenance work. It was not realized, however, that the second valve created a dead space between the two block valves and that water had been trapped between them. When the valve at the bottom of the tower was opened, the water in the piping between the valves mixed with the hot oil and flashed to steam, which surged upwards through the tower. Excessive pressures caused by this surge of steam damaged the trays.

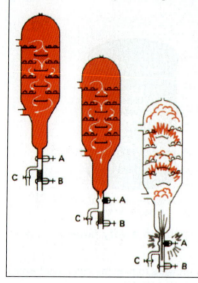

Oil temperature was raised to 260°C (500°F) with water trapped between valves 'A' and 'B'. Valve 'A' was opened and allowed hot oil and water to mix. Water flashed to steam and upset trays. Position of valve 'C' connection prevented draining.

This incident shows not only how damaging small quantities of water in vacuum towers can be, but also why the effect of any changes made in a process unit must be carefully analyzed. Thus, adding a single valve for safety created a hazardous condition when the prescribed start-up procedure was followed. *Even minor design changes must be considered carefully to determine if any change in procedure is needed to prevent hazardous operations. All process design changes, including changes to normal operating conditions, should be reviewed thoroughly by a formalized Management of Change Procedure Process (MOC).*

6.4 Hazards of atmospheric relief valves

ACCIDENT During start-up of a CDU after turnaround, water which had not been drained properly from the pumparound circuit was introduced into a hot tower. Atmospheric pressure relief valves (ARVs) located at the top of the main column opened for about 30 minutes discharging crude oil over the refinery and crude oil droplets over the adjacent neighbourhood.

(continued)

Ground fires occurred in the refinery and a major clean up of the surrounding area was required. All trays in the CDU column were damaged.

6.5 Internal VDU tower fire during turnaround

The following expands on the account contained in Section 2.2.

ACCIDENT A fire occurred at the main distillation column of a Vacuum Distillation Unit, which had been undergoing shutdown in preparation for major overhaul. The fire started in a structured packed bed inside the main distillation column, located at the bottom of the wash oil section near to the bottom of the column. The ingress of air into the main column reacted with pyrophoric scale within the beds, providing the ignition source for a large quantity of coke which had formed or migrated into the packing. Some packed sections retrieved were heavily coked.

Estimated cost of loss was $1.2 million.

| Internal damage | Bulge on shell |

6.6 Four fatalities during repair of piping

Tosco Avon Refinery, USA (Abstracted from CSB Report No. 99-014-1-CA dated March 2001)

ACCIDENT During the removal of piping attached to a 46 m (150 foot) high crude tower, naphtha was released which ignited. The flames engulfed five workers located at different heights on the tower. Four workers died and one sustained serious injuries.

A pinhole leak was discovered on the inside of the top elbow of the naphtha piping, near where it was attached to the crude tower at 34 m (112 feet) above ground. Refinery personnel responded immediately, closing four valves in an attempt to isolate the piping, while the unit remained in operation.

Subsequent inspection of the naphtha piping showed that it was extensively thinned and corroded. A decision was made to replace a large section of the naphtha line. Over the 13 days between the discovery of the leak and the fire, workers made numerous unsuccessful attempts to isolate and drain the naphtha piping. The pinhole leak reoccurred three times, and the isolation valves were re-tightened in unsuccessful efforts to isolate the piping.

Nonetheless, refinery supervisors proceeded with scheduling the line replacement while the unit was in operation. On the day of the incident, the piping contained approximately 340 litres (90 gallons) of naphtha, which was being pressurized from the running process unit through a leaking isolation valve. A work permit authorized maintenance employees to drain and remove the piping. After several unsuccessful attempts to drain the line, a maintenance supervisor directed workers to make two cuts into the piping using a pneumatic saw. After a second cut, naphtha began to leak. The supervisor directed the workers to open a flange to drain the line. As the line was being drained, naphtha was suddenly released from the open end of the piping that had been cut first. The naphtha ignited, most likely from contacting the nearby hot surfaces of the crude tower, and quickly engulfed the tower structure and personnel.

Lessons learned from this accident

The opening of plant and equipment requires:

- Careful work planning with a job hazard analysis;
- A work permitting system that:
  - assigns authority and accountability for all aspects of the work;
  - provides written confirmation that a work site is prepared and safe before work commences;
  - describes all outstanding potential hazards and required precautionary measures;
  - ensures that plant/equipment is formally handed back to operations in a safe state for recommissioning;

(continued)

- o ensures higher risk maintenance activities are authorized by a higher level of management staff; and

- o controls who does the work and at what time.

Work permit systems must be regularly monitored by management and periodically independently audited to ensure that procedures are being effectively implemented.

Permit issuing and performing authorities must be certified competent to undertake these duties following formal training.

Arrangement at the time of the initial leak. →

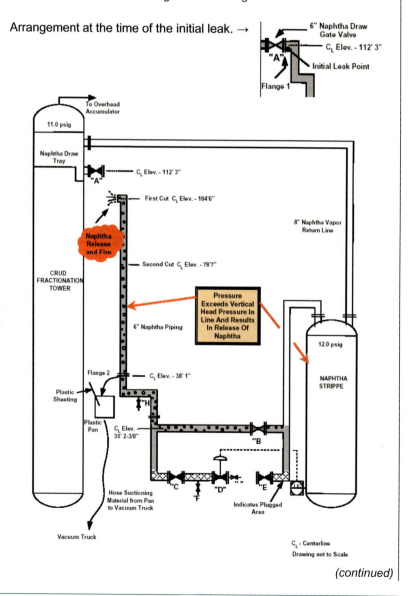

(continued)

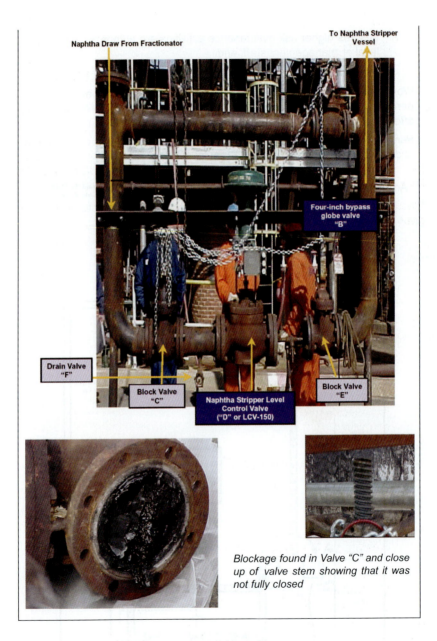

Naphtha Draw From Fractionator

To Naphtha Stripper Vessel

Four-inch bypass globe valve "B"

Drain Valve "F"

Block Valve "C"

Block Valve "E"

Naphtha Stripper Level Control Valve ("D" or LCV-150)

Blockage found in Valve "C" and close up of valve stem showing that it was not fully closed

7

Self test questionnaire

The following 21 questions are designed to assess the effectiveness of knowledge transfer following review of this booklet. It assumes a basic knowledge of plant and equipment operation.

Please note that some questions have more than one answer. Answers are given at the end of the test.

7.1 When is a process plant most vulnerable to a process safety accident:
   a) Start-up and shutdown?
   b) Normal operating period?
   c) Turnarounds?
   d) All of the time?

7.2 What are the major hazards found on CDU/VDU's:
   a) Large hydrocarbon inventories?
   b) High temperatures?
   c) High pressures?
   d) Air ingress?
   e) Water in crude oil feed?
   f) Electricity?
   g) Chemicals?
   h) All of these?

7.3 Which of the following has the lowest flash point:
   a) Naphtha?
   b) Residue?
   c) Kerosene?
   d) Gas oil?

7.4 Which of the following can form a vapour cloud on loss of containment:
   a) Residue?
   b) Butane?
   c) Cold gas oil?
   d) Hot crude oil?

7.5 What happens when a hydrocarbon is released at or above its auto-ignition temperature:
   a) Nothing?
   b) It forms a vapour cloud that drifts off-site?
   c) It ignites?
   d) It gives off a lot of smoky vapour but does not ignite?

7.6   At turnarounds, what is the greatest *process safety hazard*:
a)   Falling objects?
b)   Working at heights?
c)   Pyrophoric scale/coke fires?
d)   Noise?

7.7   When removing iron sulphide scale from equipment that has been opened up for inspection or maintenance, do you:
a)   Keep it wet until it can be removed offsite and safely disposed of?
b)   Mix it with the rest of the turnaround waste and dispose of it to landfill?
c)   Neutralize it with alkali and dispose of as a hazardous waste?
d)   Wash it down into the oily water sewer?

7.8   What is the most common type of failure mechanism that contributes to process safety accidents on CDU/VDUs:
a)   Unforeseen mechanical failure?
b)   Pump fires?
c)   Human error?
d)   Corrosion?
e)   Water freezing inside process equipment during cold weather?

7.9   If you are attempting to drain down a piece of equipment prior to maintenance, and nothing comes out when you start to open the valve, do you:
a)   Hit it with a hammer to clear any obstruction?
b)   Continue to open the drain valve until it clears?
c)   Shut the drain valve and report it to your Supervisor?
d)   Try to poke a piece of wire up the open end to clear the obstruction?

7.10   When in-situ decoking of a CDU or VDU charge heater is complete it is necessary to open the spade/blind between the charge heater and the main column. When doing this do you:
a)   Extinguish all fires in the charge heater including pilot burners?
b)   Extinguish main burners but leave the pilots alight?
c)   Ensure that the main column is depressured by opening a connection to flare?
d)   Ensure all connections between the main column and hydrocarbon sources are blinded, and maintain a slight positive pressure of nitrogen in the column?
e)   Hydrocarbon free and blind the main column before decoking is started and turn the spade/blind without having to take any special precautions?

7.11   Corrosion is a major cause of equipment and piping failure. How does an operator best contribute to corrosion control programmes?
a)   Keep a close watch on the unit and report any leaks to supervisors?
b)   Ensure that pH of aqueous streams and other simple checks are carried out in accordance with procedures and immediately report any unusual results?
c)   Know their unit's safe operating envelope?

    d)   Ensure that corrosion control chemicals are injected into process
        equipment at the correct flowrates and concentrations?

    e)   All of these?

7.12   If the desalter water injection pumps fail, what is the *first action* you
must take:

    a)   Write it up in the unit log book?

    b)   Report it to your supervisor?

    c)   Immediately valve isolate the water injection pumps?

    d)   Activate a unit emergency shutdown?

7.13   If a fired heater tube skin temperature indicator is reading top of scale,
do you immediately:

    a)   Submit a maintenance request to get it repaired?

    b)   Immediately check the individual pass flow control valve position,
        pass outlet temperature and the physical appearance of the tube?

    c)   Reduce the firing near to the affected tube skin thermocouple?

    d)   Do nothing, these things are notoriously unreliable and it will get
        fixed anyway at the next turnaround?

7.14   The VDU offgas flow to the charge heater has increased to
unprecedented levels. Do you:

    a)   Divert the offgas flow to flare?

    b)   Investigate the possibilities of a major air leak into the vacuum
        system and prepare for an emergency shutdown?

    c)   Cut unit feedrate and charge heater firing?

    d)   Turn products to slops?

    e)   Put it down to lighter than normal feedstock and challenge
        upstream process unit and tank farm operators?

7.15   Maintenance technicians have returned a hot oil duty pump to
operations with a screwed plug missing from the pump casing. Do you:

    a)   Use the plug that you found a year ago and kept in your locker for
        just this type of situation?

    b)   Call up stores to issue you with a correctly sized plug?

    c)   Reinstate the lock out and tag out (LO/TO) and call back
        maintenance to install the correct plug?

    d)   De-isolate the pump but put a warning sign on the pump that the
        plug is missing?

7.16   A hot oil pump has been warmed up and you are to commission it.
Prior to the pump starting to turn, do you:

    a)   Fully open the suction valve, crack open the discharge valve and
        close the warm-up bypass?

    b)   Fully open both suction and discharge valves and leave the
        warm-up bypass open?

    c)   Check the suction valve is fully open, crack open the discharge
        valve, and close the warm-up bypass only when the pump is at
        operating speed?

    d)   Fully open the discharge valve, crack open the suction and leave
        the warm-up bypass open?

71

7.17 A VDU bottoms pump is fitted with double mechanical seals. When attempting to start the pump after maintenance you find the seal oil return line still has a blind in it. Do you:
   a) Continue to start up the pump and request maintenance to remove the blind on the seal oil return as soon as they can?
   b) Proceed with the start-up but only to the point of warming through the pump, but do not start it?
   c) Continue with the pump start-up but keep a steam hose blowing over the seal to prevent coke accumulation?
   d) Stop the start-up procedure, reinstate LO/TO and request maintenance to remove the blind before recommencing the start-up procedure from the beginning?

7.18 You are starting up a distillation column, but the indicated level in the base is shown to be 100% of the indicated range. Do you:
   a) Make a note in the shift log to get the level indicator checked and proceed with the start-up?
   b) Proceed with the start-up. You always make the initial equipment fill to above the 100% level as you know it will come down once you get reboiler circulation started and a bit of heat on the column bottoms?
   c) Get the operators to check that level is in the range of the manually operated sight glasses and if it is proceed with the start-up?
   d) Stop the start-up and reduce the column inventory by pumping to storage until the level comes within the indicated range. Then proceed with the start-up after checking all level alarms are functioning correctly?

7.19 The unit is about to restart after a major turnaround. As a supervisor you are instructed to print off the start-up procedure. When you do, it is dated five years ago. Do you:
   a) Read it quickly and if it looks OK use it?
   b) Write in by hand any changes you feel are needed and issue to operators?
   c) Abandon the start-up and report to Managers that the start-up cannot proceed until a procedure that has been reviewed within the past 12 months and includes changes made at the turnaround is available?
   d) Immediately initiate a Pre-Start-Up Safety Review?

7.20 The planning department have obtained a parcel of crude oil at a very competitive price. This particular crude oil has never been processed at your refinery before. Do you:
   a) Quarantine the crude and immediately initiate a Management of Change review?
   b) Insist on an assay sheet before you feed the crude into your unit?
   c) Praise them for their foresight in keeping costs down in a very competitive business environment?
   d) Double up on shift operator numbers for the first 48 hours that this crude oil is run through your unit?

7.21 You are starting up a VDU and have pulled a vacuum and established cold oil circulation. To remove any residual water in the system, do you:
   a) Drain at all the usual low points?
   b) Warm up the system very slowly, maintaining a close watch on system temperatures and pressures?
   c) You don't have to worry, there won't be any water remaining anyway after pulling the vacuum?
   d) Water mixing with oil will evaporate slowly, so no special precautions are necessary?

ANSWERS TO SELF TEST QUESTIONNAIRE

1 a; 2 h; 3 a; 4 b and d; 5 c; 6 c; 7 a; 8 c; 9 c; 10 e; 11 e; 12 c; 13 b; 14 b; 15 c; 16 c; 17 d; 18 d; 19 c; 20 a; 21 b

8

References

ANSI/API 521 *Pressure-relieving and Depressuring Systems.*

API RP 571 *Damage Mechanisms Affecting Fixed Equipment in the Refining Industry.*

API RP 578 *Material Verification Program for New and Existing Alloy Piping Systems.*

API 610 *Pumps for Refinery Service.*

API Standard 617 *Axial and Centrifugal Compressors and Expander-compressors for Petroleum, Chemical and Gas Industry Services.*

ANSI/API 618 *Reciprocating Compressors for Petroleum, Chemical, and Gas Industry Services.*

UK Health & Safety Executive research report 205/1999 *Selection Criteria For The Remote Isolation Of Hazardous Inventories.*

9

Incidents list

Serial number	Year	Country	Activity	Equipment
1	1972	Australia	Maintenance—plant i/c	Electrical
2	1972	UK	Start-up	Instrumentation
3	1972	France	Normal operation	Fired heater
4	1972	Netherlands	Normal operation	Pumps
5	1972	UK	Shutdown	Fired heater
6	1972	Italy	Normal operation	Pumps
7	1972	Italy	Maintenance—plant i/c	Process piping
8	1972	?	Shutdown	Heat exchangers
9	1973	UK	Normal operation	Pumps
10	1973	Netherlands	Normal operation	Pumps
11	1973	?	Normal operation	Process piping
12	1973	UK	Normal operation	Pumps
13	1973	Netherlands	Start-up	Process piping
14	1973	Australia	Normal operation	Process piping
15	1974	?	Normal operation	Utility piping
16	1974	?	Turnaround	Desalter
17	1975	Singapore	Start-up	Process piping
18	1975	Norway	Normal operation	Pum
19	1978	France	Normal operation	Process piping
20	1978	Netherlands	Normal operation	Pump
21	1979	Netherlands	Maintenance—plant i/c	Pump
22	1979	Germany	Start-up	Heat exchanger
23	1981	Germany	Normal operation	Valve (process)
24	1982	Netherlands	Normal operation	Fired heater
25	1981	Netherlands	Shutdown	Fired heater

Serial number	Year	Country	Activity	Equipment
26	1982	?	Normal operation	Compressor
27	1982	Singapore	Normal operation	Process piping
28	1983	Belgium	Normal operation	Heat exchanger
29	1983	?	Shutdown	Process piping
30	1983	Netherlands	Start-up	Fired heater
31	1983	Australia	Normal operation	Process piping
32	1984	Australia	Start-up	Fired heater
33	1984	UK	Start-up	Pump
34	1984	Singapore	Maintenance—plant i/c	Process piping
35	1987	Netherlands	Maintenance—plant i/c	Fin-fan cooler
36	1987	Australia	Normal operation	Fired heater
37	1987	UK	Normal operation	Pump
38	1988	Australia	Start-up	Fired heater
39	1988	Netherlands	Shutdown	Columns/vessel
40	1991	USA	Start-up	Utility piping
41	1991	USA	Start-up	Fired heater
42	1993	USA	Normal operation	Process piping
43	1993	USA	Maintenance—plant i/c	Process piping
44	1994	USA	Start-up	Fired heater
45	1994	UK	Shutdown	Columns/vessel
46	1990	Middle East	Normal operation	Process piping
47	1994	UK	Maintenance—plant i/c	Pressure relief valve
48	1994	France	Turnaround	Columns/vessel
49	1994	USA	Normal operation	Pressure relief valve
50	1994	USA	Start-up	Columns/vessel
51	1994	Netherlands	Maintenance—plant i/c	Heat exchanger
52	1994	UK	Normal operation	Pump
53	1994	UK	Normal operation	Columns/vessel
54	1995	USA	Normal operation	Process piping
55	1996	UK	Normal operation	Process piping
56	1996	UK	Normal operation	Desalter

Serial number	Year	Country	Activity	Equipment
57	1995	Singapore	Normal operation	Fired heater
58	1996	UK	Turnaround	Columns/vessel
59	1996	Germany	Normal operation	Pressure relief valve
60	1996	USA	Normal operation	Columns/vessel
61	1997	USA	Normal operation	Electrical
62	1997	USA	Normal operation	Fired heater
63	1997	France	Normal operation	Pump
64	1997	Singapore	Turnaround	Columns/vessel
65	1996	USA	Normal operation	Process piping
66	1987	USA	Normal operation	Pump
67	1996	UK	Turnaround	Technical Bulletin
68	1998	Germany	Normal operation	Pump
69	1998	Netherlands	Turnaround	Process piping
70	1998	USA	Start-up	Process piping
71	1999	UK	Normal operation	Process piping
72	1998	Australia	Normal operation	Process piping
73	1999	USA	Maintenance—plant i/c	Process piping
74	2002	USA	Normal operation	Columns/vessel
75	2002	France	Normal operation	Fired heater
76	2003	UK	Maintenance—plant i/c	Fin-fan cooler
77	2004	USA	Normal operation	Process piping
78	1977	Canada	Normal operation	Process piping
79	1977	France	Normal operation	Pump
80	1987	Belgium	Normal operation	Pump
81	1999	France	Maintenance—plant i/c	Desalter
82	1991	Singapore	Start-up	Process piping
83	1993	Singapore	Maintenance—plant i/c	Process piping
84	1998	UK	Turnaround	Columns/vessel
85	1998	USA	Normal operation	Heat exchanger
86	1998	UK	Maintenance—plant i/c	Heat exchanger
87	1996	USA	Normal operation	Columns/vessel

Serial number	Year	Country	Activity	Equipment
88	2002	Australia	Normal operation	Electrical
89	2004	USA	Normal operation	Instrumentation
90	1971	France	Maintenance—plant i/c	Underground sewer
91	0	France	Normal operation	Fired heater
92	0	Germany	Normal operation	Process piping
93	1972	UK	Start-up	Process piping
94	1973	France	Maintenance—plant i/c	Fin-fan cooler
95	1973	UK	Normal operation	Pump
96	1976	?	Normal operation	Process piping
97	1978	?	Normal operation	Process piping
98	1978	?	Normal operation	Pump
99	1978	?	Normal operation	Pump
100	1979	UK	Normal operation	Pump
101	1971	France	Normal operation	Flexible hoses
102	1974	UK	Normal operation	Pump
103	1979	UK	Start-up	Pump
104	1979	Australia	Turnaround	Fired heater
105	1980	Netherlands	Normal operation	Pump
106	1980	Germany	Maintenance—plant i/c	Heat exchanger
107	1981	France	Normal operation	Process piping
108	1981	France	Maintenance—plant i/c	Instrumentation
109	1981	France	Turnaround	Fired heater
110	1982	Sierra Leone	Start-up	Instrumentation
111	1982	Sierra Leone	Maintenance—plant i/c	Instrumentation
112	1982	Canada	Start-up	Fired heater
113	1984	Singapore	Shutdown	Fired heater
114	1985	Australia	Start-up	Fired heater
115	1989	USA	Normal operation	Process piping
116	1990	?	Maintenance—plant i/c	Heat exchanger
117	1991	USA	Normal operation	Process piping
118	1991	USA	Normal operation	Pump

Serial number	Year	Country	Activity	Equipment
119	1993	USA	Start-up	Process piping
120	2001	USA	Turnaround	Columns/vessel
121	2002	UK	Maintenance—plant i/c	?
122	2006	Japan	Normal operation	Process piping
123	1996	Japan	Normal operation	Process piping
124	1997	Japan	Maintenance—plant i/c	Pump
125	1997	Japan	Maintenance—plant i/c	Instrumentation
126	1996	Japan	Normal operation	Process piping
127	1996	Japan	Normal operation	Process piping
128	1990	Japan	Normal operation	Process piping
129	1990	Japan	Maintenance—plant i/c	
130	1990	Japan	Normal operation	Valve (process)
131	1987	Japan	Turnaround	Fin-fan cooler
132	1978	Japan	Normal operation	Pump
134	2006	Lithuania	Normal operation	Process piping
135	2007	USA	?	Pump
136	2006	Japan	?	?
137	1978	France	Normal operation	Process piping
138	1979	Germany	Start-up	Valve (process)
139	2007	Australia	Normal operation	Pump drain
140	2007	Germany	Normal operation	Process piping
141	2007	Sweden	Start-up	Process Piping
142	2006	France	Normal operation	Column
143	2005	France	Normal operation	Column & RV
144	1993	France	Normal operation	Process piping

10

Glossary

AIT	Auto Ignition Temperature
ALARP	As Low As Reasonably Practicable
ARV	Atmospheric Relief Valve
BA	Breathing Apparatus
BLEVE	Boiling Liquid Expanding Vapour Explosion
BPD	Barrels Per Day
CDU	Crude Distillation Unit
EPA	Environmental Protection Agency (USA)
FBP	Final Boiling Point
FCCU	Fluid Catalytic Cracking Unit
FP	Flash Point
H_2S	Hydrogen Sulphide
HAZID	HAZard IDentification study
HAZOP	HAZard OPerability study
HIPS	High Integrity Process Shutdown systems
HP	High Pressure
HSE	Health and Safety Executive (UK)
IBC	Intermediate Bulk Container
IBP	Initial Boiling Point
IP	Institute of Petroleum (UK)
LCO	Light Cycle Oil
LEL	Lower Flammable Limit
LOPA	Layers Of Protection Analysis
LP	Low Pressure
LPG	Liquified Petroleum Gas

MSDS	Material Safety Data Sheet
NAC	Naphthenic Acid Corrosion
NORM	Normally Occurring Radioactive Materials
NRV	Non-Return Valve
OEL	Occupational Exposure Limits
P&ID	Process & Instrumentation Diagram
PORV	Pilot Operated Pressure Relief Valve
PSM	Process Safety Management
PSSR	Pre-Start Up Safety Review
PWHT	Post Weld Heat Treatment
SIL	Safety Integrity Level
SOE	Safe Operating Envelope
STEL	Short Term Exposure Limit
SWS	Sour Water Stripper
TAN	Total Acid Number
TLV	Threshold Limit Values
UFL	Upper Flammable Limit
VDU	Vacuum Distillation Unit
VLCC	Very Large Crude Carrier